# Performance Analysis of Network Architectures

Dietmar Tutsch

# Performance Analysis of Network Architectures

With 98 Figures and 12 Tables

 Springer

*Author*

Dietmar Tutsch

Institut für Technische Informatik
und Mikroelektronik
Technische Universität Berlin
Einsteinufer 17
10587 Berlin
Germany
DietmarT@cs.tu-berlin.de

ISBN   978-3-642-07067-9          e-ISBN   978-3-540-34310-3

ACM Computing Classification (1998): C.2, C.4, G.3, I.6

Springer is a part of Springer Science+Business Media

springer.com

© Springer-Verlag Berlin Heidelberg 2010
Printed in Germany

Cover design: KünkelLopka Werbeagentur, Heidelberg

# Preface

This monograph has only become possible due to broad support from many people. Without their work and ideas, the research presented would never have been as successful.

I would like to thank Prof. Günter Hommel for supervising my research and for his valuable comments and suggestions, which were very important for this work. Prof. Hommel has always been addressable for fruitful discussions. In addition, I am deeply grateful to Prof. Miroslaw Malek for his continuous support of my work. He has spent much time in meetings with me to direct my interest to several new topics.

This monograph also represents the author's habilitation thesis. Sincere thanks goes to all its reviewers, Prof. Hommel and Prof. Malek as well as Prof. Manfred Glesner and Prof. Peter Kropf. In addition, I am indebted to Prof. Hans-Ulrich Heiß and Prof. Thomas Sikora who joined the habilitation committee with Prof. Hommel and Prof. Malek. I appreciate their time and effort for supporting my work.

Furthermore, I am very grateful to Joachim Sokol for supporting my research on ubiquitous computing. I would also like to thank all my former and current colleagues for discussions and resulting ideas and concepts. Particularly, I am very grateful to Daniel Lüdtke and Marcus Brenner for our weekly meetings and discussions about our common work on network architectures. Sincere thanks also go to Eveline Homberg for drawing many figures of this monograph.

Finally, I am indebted to all those students of TU Berlin who have been involved in my research, especially to Daniel Benecke, Rainer Holl-Biniasz, Matthias Kühm, and Arvid Walter, who implemented many components in our software tools and did excellent work.

Technische Universität Berlin,

*Dietmar Tutsch*
June 2006

# Contents

# 1

# Introduction

Parallel and distributed computer systems are built to close the gap between high-performance demands in computing and the available computing power of stand-alone single processor machines. Choosing an appropriate network architecture to connect the parallel or distributed computer system plays an important role in the development process of the system. An ineligible network architecture may significantly delay the communication between the parallel or distributed components and decrease the system performance.

This book introduces the different kinds of network architectures. Wired architectures as well as wireless ones are taken into account. Various network topologies are presented, including their main components, the switches which realize the input-output connections in network nodes. The internal structure of networks and switches is exhaustively discussed. In addition, a new network topology is presented.

The advantages and drawbacks of the various switch structures and network topologies are described to work out their favored application areas. For instance, these areas depend on their related network traffic. Thus, the network traffic is also focused on in this book. The traffic is characterized by its distribution in time and space. Knowing the distribution in time is important to sufficiently dimension the network components, e.g., the buffer sizes. On the other hand, knowing the distribution in space helps choose an appropriate network topology supporting this distribution.

An important item in characterizing the advantages and drawbacks of switch and network architectures is comparing their performance. To improve this characterization by performance evaluation, a new method to determine network performance of parallel and distributed systems is derived in this book. Furthermore, guidelines and engineering techniques are given to rapidly establish models of the system in question, to reduce their complexity, and to solve them in a time-saving manner. These guidelines result in a systematization of the model development process and help set up appropriate models.

## 1.1 Motivation

To evaluate the performance of networks is a challenging task. Many parameters influence the performance of a particular network architecture. To determine its performance, three methods can be applied.

1. Measurement: The system hardware has to be set up to connect measurement devices.
2. Simulation: A software model of the system has to be set up, and simulation runs lead to the performance results.
3. Mathematical methods: Systems of equations have to be established to model the system in question. Performance is derived by solving these equations.

Measurement suffers from the huge drawback that the system must first be realized in hardware before any measurement can start. If it then turns out that the system does not fulfill the required performance, the architecture must be changed. Repetitively reconstructing it several times until the optimal architecture is found consumes much time and money. Thus, modeling with simulation or mathematical methods is the most appropriate technique. Nevertheless, measurement should not completely be excluded. Many systems are too complex to be determined by simulation or mathematical methods if detailed and accurate results are required. Then, measurement often turns out to be the only feasible solution. Furthermore, measurement also helps validate any results achieved by simulation or mathematical methods. Inaccuracies and errors in those models can be detected.

This book mainly focuses on simulation and mathematical methods. Features, strengths, and drawbacks related to simulation and mathematical methods are investigated. But before these techniques can be applied to evaluate parallel and distributed system architectures, models according to the methods must be derived.

Model development is a difficult task. Experience is needed, as well as creativity. Usually, no general rules can be given for model development. But guidelines can be presented from the experience acquired while establishing models. One should be aware that badly engineered models lead to time-consuming performance evaluation, or they are not even able to produce any results due to their huge size and limited computer power and memory. Therefore, this book additionally introduces model engineering techniques related to the investigated performance evaluation methods.

Summarizing previous items, the following challenges emerge:

- Fast model development is desired. Guidelines of experienced developers may help in model set-up. These guidelines must be established.
- Models should be developed in such a way that they can be simulated or mathematically solved in a short time to produce results fast. Guidelines and model engineering techniques that lead to such models are required.

- Models should accurately represent the real-world system. Previously mentioned guidelines and engineering techniques should consider this.

The following chapters give some guidelines and model engineering techniques that fulfill the above issues and help in model development.

## 1.2 Contribution

This book focuses on the model engineering of network and switch architectures for parallel and distributed systems. Particularly, simulation models and mathematical models are of interest. The models are established to allow system performance evaluation. Examining the architecture implies coping with different network topologies, different buffer positions and sizes, different switch sizes, etc. Routing, protocols, and fault tolerance are not in the scope of this book.

In Chap. 2, the characteristics of networks for parallel and distributed systems are described, including switching techniques, traffic patterns, and wired and wireless network architectures. With regard to traffic patterns, not only if the distribution in time taken into account, the distribution in space is also considered. Particularly, multicast distributions and the network topology support for multicast are examined. As a result of this research, a new wired network architecture was developed, and is presented in Sect. 2.3.10. This new architecture is called multilayer multistage interconnection network (MLMIN), and heavily increases network performance in the case of multicast traffic.

Methods for performance evaluation of parallel and distributed systems are discussed in Chap. 3. Simulation and mathematical methods are presented. Simulation refers to discrete event simulation only. Continuous simulation is not taken into account due to its drawbacks (see Sect. 3.1). With regard to mathematical methods, the focus is on Markov chains and Petri nets. Queuing theory suffers from its problems in modeling complex network topologies and is thus not considered here.

Establishing a simulation model or a mathematical model calls for some "feeling" of how to design an appropriate model. A model quickly becomes too large or too complex, resulting either in huge calculation times or, worse, in no solution at all. On the other hand, setting up a small model with low calculation times may consume much time during development. Furthermore, small models usually neglect too many details, and are thus too inaccurate.

In Chap. 4, some new guidelines for model development and complexity reduction are given. These guidelines evolved while establishing many models of communication networks to determine their performance.

The presented guidelines include a new concept of automatic model generation that gives a strategy for how to develop a generator for automatic model derivation. Of course, such a concept saves much development time.

Two examples show how to apply the above concepts. A smaller example presented in Chap. 5 deals with a cellular network. It examines the handoff procedure of mobile nodes carrying real-time traffic. Due to the modest model size, a Petri net description sufficiently handles the system in question.

Unfortunately, this is not valid for the second example, which is dealt with in Chap. 6. Multistage interconnection networks are modeled to optimize their architecture. They are of common interest due to their use in parallel computers and in switches connecting distributed systems; their reconfiguration properties are also examined currently [124].

It turns out that this system is too large to be simply modeled by Petri nets. Other techniques like simulation and Markov chains are also applied and compared. The automatic model generation particularly accelerates model development, and is thus a good option in model establishment.

## 1.3 Related Work

Much research has been performed in investigating the performance of parallel and distributed systems, particularly in the area of communication networks connecting the nodes of such systems. Methods to evaluate the performance include Markov chains, queuing theory, Petri nets, and simulation (see Chap. 3). The most important publications are referenced in the following.

A broad theory about Markov chains is provided by many publications [4, 19, 44, 53, 54, 92, 95, 96, 199]. Especially, [19] does not introduce only Markov chains: The research groups of the authors also focus on queuing theory. Thus, this book also states the relation between Markov chains and queuing theory. Other publications introducing queuing theory are, for instance, [61, 72, 95, 96, 137, 199].

Some research groups which deal with Markov chains also apply Petri nets as a modeling method. Their publications and those of other authors describing Petri nets are, for instance, [36, 63, 75, 86, 118, 128, 144, 160, 165]. Finally, simulation methods are exhaustively treated in [9, 23, 57, 88, 107, 248].

Some publications particularly deal with the performance evaluation of multistage interconnection networks, which is one of the applications of this book. In the following, it is concentrated on research that was published during the last years. Older publications (but also newer ones) are cited in the chapters of this book to which they are related, where they are discussed in detail.

Group communication in circuit-switched multistage interconnection networks (MINs) is investigated in [242] by applying Markov chains. Markov chains are also used in [250] to compare MIN performance in the case of different buffering schemes. Hot-spot traffic performance in MINs is examined in [87]. [200] deals with multicast in Clos networks as a subclass of MINs. One of the authors of the previously mentioned paper also published [235], where

MINs are used to establish active routers. Further publications describe multistage interconnection networks that connect a parallel computer [191] and MINs used to set up ATM switches [184].

The performance evaluation of wireless networks is broadly based [150, 174, 185, 190]. Some publications relate to handoffs in a cellular network, which is close to one of the applications of this book. For instance, the book deals with the spatial distribution of wireless networks and their performance evaluation using Markov chains [25]. Further publications evaluating performance in the case of handoffs and considering the influence of architectures are [12] (using Markov chains) and [132] (using Petri nets). Markov chains as a method to model a cellular system supporting traffic of multiple bandwidth requirements are also applied in [101, 131]. Some assumptions about the movement of mobile nodes are given in [29].

## 1.4 Term Definitions

The basic terms of this book are introduced below, before the following chapters will use them in the defined context. Unfortunately, terminology in this field is not unique. Thus, other books may define them differently.

### 1.4.1 Models and Performance Evaluation

The term "performance evaluation" covers all kinds of methods to determine system performance. Section 1.1 already enumerated and classified these methods as measurement, simulation, and mathematical methods. This book mainly concentrates on the last two items. A detailed description of them will be given in Chap. 3. The methods are applied to evaluate network architectures for parallel and distributed systems.

The performance to be determined is usually described by latency and throughput of the network or of a part of it. Latency defines the time between the initiation of a message transmission at the source and its complete receipt at the destination [51]. That means the latency is the sum of the time for communication overhead at the source, signal delay time, ratio of message length to bandwidth, and the time for communication overhead at the destination [163]. Investigating network architectures, the communication overheads at the source and destination are usually not of interest. Furthermore, the ratio of message length to bandwidth can be neglected if message units are very small. Then, the delay time becomes the only quantity to characterize the latency (e.g., [32, 244]), particularly if no direct connection between the source and the destination is established. Comparing several network architectures, the delay times are compared by global clock cycles rather than by absolute values of time units [51]. This book also characterizes network architectures by their delay times measured in clock cycles.

The throughput of a network is defined as the amount of information delivered per time unit [51]. Again, this quantity can be determined related to absolute values of time units or to clock cycles, as practiced in the remaining part of this book. The throughput heavily depends on network size. Thus, the throughput is usually normalized by dividing it by the network size.

Simulation or mathematical methods for performance evaluation are applied by modeling the system in question. A model is an abstract or theoretical representation of this system. Many different models can be used to describe the system behavior. Often, a simplified system behavior is modeled to keep the model size reasonable.

### 1.4.2 Parallel System Architecture

A parallel system refers to a system that executes an action simultaneously with tightly coupled devices [194]. The devices support each other via the exchange of intermediate results received by performing a part of the action. If a device fails, the action usually cannot be completed, which especially corresponds to failed connections between the devices.

Tanenbaum [196] defines a parallel system (for instance, a multicomputer system) as a system that "(simultaneously) works on a single problem."

Parallel systems can be found in many areas. This book focuses on parallel systems related to computer science: devices (e.g., nodes of a computer system) are connected by a communication network and perform an action in close cooperation [1, 2, 115, 119, 167].

Multiplying matrices on a parallel computer system is a typical example. The parallel computer system consists of hundreds of processors. They are connected by a high-performance communication network, e.g., a crossbar or a hypercube (see Sect. 2.3). The matrix multiplication is divided into multiple independent scalar multiplications. Each processor performs a part of those scalar multiplications. The results are transferred to some of the processors and further evaluated there. If a processor fails (and no redundancy is available), some results are missing and the product of the matrices cannot be determined.

Parallel system architecture refers to overall features concerning the parallel system design. Its main issues are the structure and the behavior of the system.

For instance, Fig. 1.1 roughly shows the architecture of a parallel computer system. The main elements of the system are the nodes. Usually, they consist of processors and some local node memory. The nodes are all close together, e.g., on a computer main board or on separate boards that are directly connected.

Tightly coupling the devices requires in general a structure of closely located devices: a parallel system is established in a limited area.

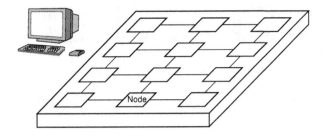

**Fig. 1.1.** Architecture of a parallel computer system

### 1.4.3 Distributed System Architecture

A distributed system refers to a system that executes an action with loosely coupled devices [194]. Communication time between the devices only represents a small fraction of the entire execution time. Therefore, the devices need not to be highly meshed. They may be distributed in a wide area.

If a device fails, distinct tasks are no longer available. But all other devices proceed with their actions. Even if the connection between them fails, distinct tasks are no longer available but all devices remain in service as stand-alone devices.

As in case of parallel systems, this book only focuses on distributed systems related to computer science: devices (e.g., stand-alone computers) are connected by a communication network and perform an action in loose cooperation (by communicating every now and then).

Distributed system architecture refers to all features concerning distributed system design. For instance, Fig. 1.2 roughly shows the architecture of a distributed computer system. It consists of several computers connected via a network. The computers commonly solve a task or can act as stand-alone machines. Even computers of distinct networks can work together if networks are connected (e.g., via a gateway or switch [6, 77]). Due to the networks, large distances may exist between computers of a distributed system [247].

Several definitions describe parallel systems as a subset of distributed systems. For instance, Tanenbaum [195, 196] defines a distributed system as "a set of independent computers that appear to the user as a single system." Discussing the hardware concept of such systems, he explicitly includes shared memory systems (multiprocessor systems) besides distributed memory systems (multicomputer systems).

In this book, parallel systems like shared memory systems are distinguished from distributed systems by their tight coupling via the memory: if a processor accesses (a part of) the memory that is located at a particular port, all other processors are blocked in the case of accessing the same (part of the) memory. As a result, communication time increases (if the delay until the data transfer between processor and memory starts is also included in the communication

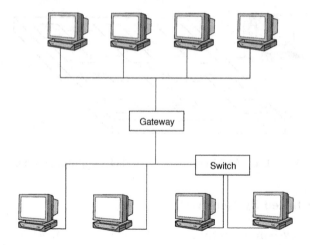

**Fig. 1.2.** Architecture of a distributed computer system

time). Communication time represents in this context the time from a data request to the memory until the data is delivered to the requesting processor.

Nevertheless, there is a smooth transition between parallel and distributed systems. For instance, if the shared memory system consists of a shared memory with multiple ports or if each processor holds its own cache memory, the processors are no longer coupled as tightly as in the previous example.

Coulouris' definition [40] of a distributed system does not agree with Tanenbaum's. Shared memory systems (with common memory) are excluded from distributed systems. He defines a distributed system as "one in which components located at networked computers communicate and coordinate their actions only by passing messages." This means that only message-passing systems belong to the set of distributed systems.

Coulouris derives from his definition three main features of distributed systems: concurrency of components, lack of a global clock, and independent failures of components. The last two features particularly emphasize the character of loosely coupled devices in distributed systems. Systems synchronized by a global clock or in which failed components cause further failures must be called tightly coupled.

### 1.4.4 Network Architecture

Previous sections show that a major component of parallel and distributed systems is their network. It allows communication between parallel or distributed devices. Due to tightly coupled devices of parallel systems, messages that are exchanged in such systems should be received with only a short delay. To meet the demands of high network performance, the devices usually are located close together, avoiding large distances to be traveled by messages within the network. As a result, networks of parallel systems only cover a local area.

The network architecture describes the network topology and its physical realization by determining the kinds and parameters of the network elements in detail. The network topology only gives the structure of the connections between the nodes related to graph theory.

Devices of distributed systems may be spread over a large area, which means that a network linking the devices is also spread over a large area. The local parts of such a network are usually connected by switches. The switches often consist of similar architectures as networks for parallel systems [51, 184]. This book focuses on such architectures.

Distributed systems that include mobile devices need networks that deal with this additional issue. Such wireless networks are also a topic of this book. The performance and dynamic behavior of links between wireless devices and wired devices are the main focus. Distributed computer systems with mobile devices allow mobile computing, also called nomadic computing [94, 126]. Ubiquitous computing [68, 161, 228] overlaps with mobile computing: ubiquitous computing assumes many computing devices in the user's environment allowing ubiquitous computer access. Such computing devices may include mobile devices. Ubiquitous computing is also called pervasive computing.

In general, devices of a parallel or distributed system, which are connected via a network, are also called communication partners or nodes in this context, independently of whether they are mobile or fixed devices.

# 2

# Characteristics of Network Architectures

In comparing parallel and distributed systems, many characteristics surface that both systems have in common. This chapter introduces those characteristics. It concentrates on characteristics highlighted by the communication aspects of parallel and distributed systems. Due to coupled devices, both kinds of systems require a network between the devices allowing communication. Common characteristics concerning message transfer (switching techniques), network architectures, and network traffic patterns are described. Protocols and fault tolerance are not investigated in this book, and are thus not addressed.

## 2.1 Switching Techniques

A common task of parallel systems and distributed systems is their need for a communication system. It connects the devices (nodes) of the system and enables any required message exchange between them. The switching techniques [45, 163, 181] describe when and how message components are transferred along a network link. They are not in charge of determining the path that the message takes through the network. This job is performed by the routing algorithm and is not subject of this book. Only some routing basics related to particular network architectures are explained in Sects. 2.3 and 2.4.

### 2.1.1 Circuit Switching

The circuit switching technique [108] reserves a physical link between the sender (source node) and the receiver (destination node). Reservation is established by a control information (probe) between source and destination. Then, the physical link is reserved for the total transfer time. Therefore, network resources may heavily be wasted if the amount of transferred data per time unit (called data rate) is lower than the rate the link is able to deal with. Furthermore, messages of other communication partners may not be exchanged due to these reserved links (message blocking).

On the other hand, after establishing a connection, no further signaling is needed during message transfer except for the releasing of the link at the end of the transfer. Thus, circuit switching performs well if messages are long.

At the hardware level, messages are usually divided into phits (physical units). A phit represents the smallest unit that can be transferred during a hardware clock cycle. Depending on the link width, a phit usually consists of at least one and up to 64 bits.

### 2.1.2 Packet Switching

The packet switching technique divides a message into several packets. Packets consist of a header field and a payload field (Fig. 2.1, top). The header

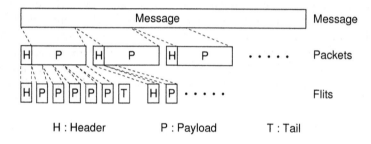

**Fig. 2.1.** Dividing a message into packets and flits

contains the destination information. Further information may also be added, like packet length (if the packets are not of equal length), sender information, or any other switching information. The payload field contains message data. Packet switching has the advantage that packets can be stored (buffered) during their transfer within the network (e.g., at intermediate nodes that have to be passed) due to their limited size. This means packets can be forwarded within parts of the network and stored at their current position in the case of blocking. Blocking may arise due to occupied buffers and links of intermediate nodes. After blocking, forwarding of buffered packets can be resumed.

Logical links are introduced describing the message path through the network. The feature of storing packets allows multiple logical links to use the same physical links: the resources are better utilized.

Three kinds of packet switching techniques are distinguished: store-and-forward switching, cut-through switching, and wormhole switching.

## Store-and-Forward Switching

In store-and-forward switching [49], packets are forwarded from the source node to the destination node while being buffered at each intermediate node. A node forwards a packet when the succeeding node is available. A move from one node to the next one is called a hop.

## Cut-Through Switching

In cut-through switching [91, 179, 209, 230], packets are forwarded from the source node to the destination node in a manner similar to store-and-forward switching. But packets are buffered at an intermediate node only if the succeeding node is not available. In other words, packets move through ("cut through") the network by passing intermediate nodes until they are blocked. On the hardware level, the phits are forwarded in a pipeline manner.

If blocking occurs, packets are buffered until the blocking is released: virtual cut-through switching blocks the packet until the last phit arrives at this buffer, while in partial cut-through switching, phits are already allowed to proceed even if not all phits have reached the buffer where the packet is blocked.

Due to less buffering, packets traverse a network much faster than in the case of store-and-forward switching if only few blockings occur. If blocking occurs at each intermediate node, cut-through switching performs like store-and-forward switching.

## Wormhole Switching

In wormhole switching [21, 87, 136, 139, 148], packets are further divided into logical units called flits (flow control units). All flits (Fig. 2.1, bottom) are of equal size (usually between one and eight bytes). The first flit contains the header information and reserves a link through the network while it is forwarded in a cut-through switching manner. The following flits contain the payload. The last flit (called tail) deallocates the link.

Each intermediate node has a buffer of flit size. As in the case of cut-through switching, the header flit moves through the network by passing intermediate nodes until it is blocked. If blocking occurs, the header flit is buffered at the related intermediate node. All other flits belonging to the same packet are buffered at their current intermediate nodes.

Dividing packets into flits of equal size reduces buffer costs. Intermediate nodes only need buffers of the flit size. On the other hand, if flits are blocked and buffered, the packet is spread over a part of the network. Due to the reserved link between the header flit and the tail flit, such a packet blocks a part of the network.

## 2.2 Traffic Patterns

Besides switching techniques, parallel and distributed systems also show some
other common characteristics concerning their network traffic patterns. A particular network traffic pattern results from the communication requests of the
network nodes. The times at which messages are sent and their particular
destinations determine the traffic of one node. Superposing the traffic of all
nodes gives the traffic pattern of the network.

That means the network traffic patterns are defined by the distribution of
messages in space as well as by their distribution in time.

### 2.2.1 Distribution in Space

The distribution of messages in space can be related to the varying message
density in network areas. But it can also describe the varying number of nodes
a message is destined to. Both aspects will be discussed here.

**Uniformity and Hot Spot**

Ideally, the message density is uniformly distributed all over the network.
Such a traffic pattern ensures an equal load on all communication links. This
is particularly important if all links have equal capacity.

Therefore, uniform traffic allows the design of very regular network structures, simplifying the network design. Equal network components can be used
independently of the particular network area that is considered. Furthermore,
establishing mathematical models of the network becomes much easier due to
the regular and symmetric structure and uniform traffic, as will be seen in
Sect. 3.2.

On the other hand, tasks or devices may be asymmetrically distributed
among the network nodes. That means nodes are not equal in either their
hardware or their software. Then, messages are usually non-uniformly distributed in the network. For instance, if only one node owns a main memory
to store a huge amount of data, all other nodes send their data to this memory:
a very high load results on the link to the memory node. In such a scenario,
this link is called a hot spot [5, 100, 110, 166], and it may become a bottleneck
of the system. More generally, a hot spot is any device in a system that turns
into a performance bottleneck due to high utilization.

**Multicast**

The varying number of nodes a message is destined to also describes a distribution in space. Sending a message to multiple destination nodes is called
multicast [151, 182, 192, 200, 207, 209, 212, 225, 241]. The term multicast
includes the two particular cases of unicast and broadcast. Unicast denotes

the process of sending a message to only a single destination node whereas broadcast represents the process of sending a message to all destination nodes of the network.

A multicast can be realized either by message replication before routing (MRBR) or by message replication while routing (MRWR) [74, 238]. In MRBR [239], a multicast message that is destined to $i$ nodes is copied $i$ times before it is routed through the network. Then, those $i$ messages are treated as unicast messages, each of them sent to one of the required destinations. In MRWR [18], the multicast message is sent as one message into the network. It proceeds through the network until it reaches a node or switch from which no single way leads to all required destinations. There, the message is copied as many times as there are different ways to reach all desired destinations. Such nodes or switches may be passed several times.

### 2.2.2 Distribution in Time

The distribution of messages in time addresses the time-dependent variation of the traffic density.

#### Initial Transient Phase

The simplest case gives a scenario in which the traffic density changes only for a limited time interval and then ends up in a steady state. The network is said to be in an initial transient phase during the time of traffic change.

For instance, such a scenario arises if an idle network is connected to nodes that are sending messages at a constant rate. Depending on the network size and the rate, the network needs some time until it is filled with messages and remains in a steady state with regard to the quantity of messages in the network.

#### Basic Functions of Time

The shape of time-dependent traffic in networks can be described by functions of time. Such a function of time that represents a single message in the network is also called signal in telecommunications. Signals that show a very simple shape are said to be basic signals. Table 2.1 introduces some basic signals. Symbol $t$ denotes the time. Replacing $t$ by $t - t_0$ allows a time shift of the function by time $t_0$. Replacing $t$ by $\frac{t}{T}$ stretches the function by the constant $T$.

Usually, time-dependent traffic [204] in networks consists of a superposition of many signals sent from many source nodes. Particularly, the shape of such a superposition describes the distribution of messages in time (time-dependent variation of the traffic density).

Many distributions in time that show specific behavior are distinguished. Two of them, which result in an often observed traffic density shape, are addressed below: bursty traffic and multifractal traffic.

**Table 2.1.** Basic signals

| Signal | Description |
|---|---|
| Sinus | $$s(t) = \sin(2\pi t)$$ |
| Gauss | $$s(t) = e^{-\pi t^2}$$ |
| Heaviside | $$\varepsilon(t) = \begin{cases} 0 & : \quad t < 0 \\ 1 & : \quad t \geq 0 \end{cases}$$ |
| Rectangular | $$\text{rect}(t) = \begin{cases} 1 & : \quad |t| < \frac{1}{2} \\ 0 & : \quad |t| \geq \frac{1}{2} \end{cases}$$ |
| Dirac | $$\delta(t) = \lim_{T_0 \to \infty} \frac{1}{T_0} \text{rect}\left(\frac{t}{T_0}\right)$$ |

**Bursty Traffic**

A special kind of time-dependent traffic is called bursty traffic [58, 205]. Bursts denote a high amount of transferred messages in a very short time. Bursty traffic consists of bursts separated by a longer period of fewer messages. Figure 2.2 gives an example. In parallel and distributed systems, bursts may be caused by transferring large files, large databases, large program code, and so on.

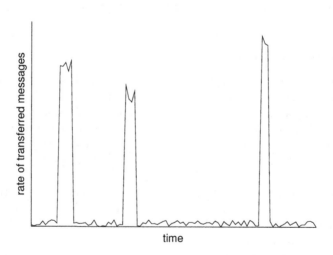

**Fig. 2.2.** Bursty traffic

**Multifractal Traffic**

To define multifractal traffic [24, 28, 52, 215], self-similar traffic is introduced first. Self-similar traffic is defined as a traffic pattern that is invariant against changes in scale or size [116, 121, 122, 186, 232]. If a part of the self-similar traffic is cut out and magnified, it will show the same structure and behavior as the non-magnified original traffic.

An approximation of a self-similar traffic is depicted in Fig. 2.3. It is a

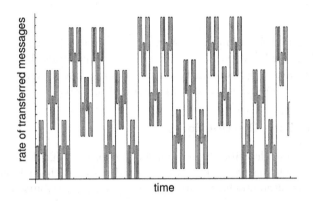

**Fig. 2.3.** Approximation of self-similar traffic

kind of a rectangular function. If a part of this function is cut out and magnified, this function again shows the same structure. This can be performed for several steps (time scalings). This self-similarity arises because the function of Fig. 2.3 is built by superposing periodical rectangle functions of different frequencies. The different frequencies represent different time scale factors.

Figure 2.3 does not really represent self-similar traffic because cutting out and magnifying the function for a few time scalings yields a similar structure, but the similarity stops if scaling is continued.

Self-similarity can easily be defined if discrete time traffic is assumed. For instance, this is true for clocked systems. A time discrete signal $x(t)$ is said to be self-similar with parameter $\beta$ ($0 < \beta < 1$) if, for all $m \in \mathbb{N}\backslash\{0\}$,

$$\mathrm{Var}(x^{(m)}) = \frac{\mathrm{Var}(x)}{m^{\beta}} \quad \text{and} \tag{2.1}$$

$$\rho_{x^{(m)}}(t) = \rho_x(t) \tag{2.2}$$

hold, where Var denotes the variance, $\rho$ the autocorrelation, and

$$x^{(m)}(\tau) = \frac{1}{m} \sum_{t=\tau m-(m-1)}^{\tau m} x(t) \tag{2.3}$$

describes the average over $m$ values of the original signal. Section 3.1.1 deals with the calculation of the variance and the autocorrelation.

The parameter $\beta$ indicates the self-similarity. It is related to the Hurst parameter $H$ which gives the degree of long-range dependence:

$$H = 1 - \frac{\beta}{2}. \tag{2.4}$$

A value of $H \leq 0.5$ ($\beta \geq 1$) expresses the lack of self-similarity. The closer $H$ is to 1, the greater the self-similarity. $H = 1$ means $\beta = 0$, and results in

$$\mathrm{Var}(x^{(m)}) = \mathrm{Var}(x). \tag{2.5}$$

If $\beta$ is not constant for all $m$, but constant with value $\hat{\beta}$ for small $m$, and with a slight change for larger $m$, such traffic is called multifractal.

The definition of self-similarity for continuous time traffic emerges from an extension of above theory, and can be found in [186].

### Interarrival Time Distribution

The above sections describe the temporal behavior of traffic density in several ways. When modeling parallel and distributed systems, the behavior in time of the traffic is often stochastically described by the distribution of the time passing by until an event occurs. Such an event may be the random arrival of a new packet at a communication network. Then, this distribution is called the interarrival time distribution of the new packets. Waiting times, service times, and so on are also described by the same distributions.

The distribution function $A(t)$ denotes the probability that the event under consideration occurred in the interval $[0, t]$. This function is also called the cumulative distribution function (CDF). Its derivative

$$a(t) = \frac{dA(t)}{dt} \tag{2.6}$$

is the probability density function (pdf).

One of the most important distributions is the exponential distribution. Its cumulative distribution function and probability density function are given, respectively, by

$$A(t) = 1 - e^{-\lambda t} \quad \text{and} \tag{2.7}$$
$$a(t) = \lambda e^{-\lambda t}, \tag{2.8}$$

where $\lambda$ represents the mean arrival rate. If service times are described, $\lambda$ is replaced by $\mu$. The memoryless property of the exponential distribution is often used (see Sect. 3.2.1). It says that the probability of the occurrence of the investigated event in interval $[0, t]$ is equal to the probability of its occurrence in interval $[\tau, t + \tau]$ if it did not take place until time $\tau$.

The Erlang-$k$ distribution describes the sum of $k$ exponentially distributed phases with mean rate $k\lambda$. The CDF and pdf result in

$$A(t) = 1 - e^{-k\lambda t} \cdot \sum_{i=0}^{k-1} \frac{(k\lambda t)^i}{i!} \quad \text{and} \qquad (2.9)$$

$$a(t) = \frac{(k\lambda)^k \cdot t^{k-1} \cdot e^{-k\lambda t}}{(k-1)!}, \qquad (2.10)$$

with $k \in \mathbb{N}$. A generalization of this distribution is obtained if the restriction on $k$ to be a natural number is relaxed. Such a distribution is called gamma distribution, where $k$ is replaced by $\alpha$ (with $\alpha > 0$):

$$A(t) = \int_0^t \frac{\alpha\lambda \cdot (\alpha\lambda\tau)^{\alpha-1}}{\Gamma(\alpha)} \cdot e^{-\alpha\lambda\tau} d\tau. \qquad (2.11)$$

$\Gamma(\alpha)$ is defined by $\Gamma(\alpha) = \int_0^\infty \tau^{\alpha-1} e^{-\lambda} d\tau$. The Erlang-$k$ distribution results from the gamma distribution for a positive integer $\alpha = k$.

The hyperexponential distribution describes the choice between $k$ exponentially distributed phases with mean rates $\lambda_i$. The CDF and pdf are

$$A(t) = \sum_{i=1}^{k} q_i \left(1 - e^{-\lambda_i t}\right) = 1 - \sum_{i=1}^{k} q_i e^{-\lambda_i t} \quad \text{and} \qquad (2.12)$$

$$a(t) = \sum_{i=1}^{k} q_i \lambda_i e^{-\lambda_i t}, \qquad (2.13)$$

where $q_i$ represents the probability that the $i$th phase is chosen. Thus, $\sum_{i=1}^{k} q_i = 1$ holds.

To describe system failures, the Weibull distribution is often applied. Its CDF and pdf are

$$A(t) = 1 - e^{-(\lambda t)^\alpha} \quad \text{and} \qquad (2.14)$$

$$a(t) = \alpha\lambda^\alpha t^{\alpha-1} e^{-(\lambda t)^\alpha}, \qquad (2.15)$$

where $\alpha$ represents the shape parameter with $\alpha > 0$.

A distribution describing non-stochastic behavior is given by the deterministic distribution. The investigated event is assumed to take place at time $t_0$. Thus, no stochastic behavior is involved. The cumulative distribution function and probability density function result in

$$A(t) = \begin{cases} 0 & : \quad t < t_0 \\ 1 & : \quad t \geq t_0 \end{cases} \quad \text{and} \qquad (2.16)$$

$$a(t) = \begin{cases} 1 & : \quad t = t_0 \\ 0 & : \quad \text{otherwise.} \end{cases} \qquad (2.17)$$

To model multifractal traffic as presented in Sect. 2.2.2, the Pareto distribution is often proposed to describe the related interarrival times. It belongs to the group of heavy tailed distributions where the tail of the distribution is hyperbolic (i.e. $t \to \infty$). The CDF and pdf of the Pareto distribution are given by

$$A(t) = 1 - \left(\frac{b}{t}\right)^{\alpha} \quad \text{and} \tag{2.18}$$

$$a(t) = \alpha \cdot \frac{b^{\alpha}}{t^{\alpha+1}}, \tag{2.19}$$

with $0 < \alpha < 2$. It is defined over the interval $t \geq b$.

A distribution where each value in an interval $[b_l, b_r]$ will be chosen with equal probability is called uniform distribution:

$$A(t) = \begin{cases} 0 & : \quad t < b_l \\ \frac{t-b_l}{b_r-b_l} & : \quad b_l \leq t \leq b_r \quad \text{and} \\ 1 & : \quad t > b_r \end{cases} \tag{2.20}$$

$$a(t) = \begin{cases} \frac{1}{b_r-b_l} & : \quad b_l \leq t \leq b_r \\ 0 & : \quad \text{otherwise} \end{cases} \tag{2.21}$$

represent the its CDF and pdf.

Errors or values given by a sum of many other values are often distributed according to the normal distribution (also called Gaussian distribution). Its probability density function is given by

$$a(t) = \frac{1}{\sqrt{2\pi\sigma^2}} e^{-\frac{(t-\mu)^2}{2\sigma^2}}, \tag{2.22}$$

with the location parameter $\mu$ representing the mean and with the scale parameter $\sigma$ representing the square root of the variance (see Sect. 3.1). No closed-form cumulative distribution function for the normal distribution exists.

Many further distributions are known, e.g., Cox distribution, double exponential distribution (also called Laplace distribution), Student's $t$-distribution, lognormal distribution, and beta distribution, among others[107].

All previously mentioned distributions are continuous. Interarrival times of discrete time systems are modeled by discrete distributions. They are characterized by their cumulative distribution functions and their probability mass functions. The probability mass function $p(t)$ denotes the probability with which an arrival occurs at discrete time $t$.

Many discrete distributions exist. For instance, the geometric distribution offers the memoryless property as the exponential distribution does in the case of continuous time. The cumulative distribution function and the probability mass function are defined by

$$A(t) = 1 - (1-p)^{t+1} \quad \text{and} \tag{2.23}$$

$$p(t) = p \cdot (1-p)^t, \tag{2.24}$$

with $t \in \mathbb{N}$ represents the $t$-th time step and $(1-p)$ gives the probability that no arrival occurred in a time step.

The discrete uniform distribution describes several possible times $t_i$ at which an arrival may occur. The cumulative distribution function and the probability mass function are given by

$$A(t) = \begin{cases} 0 & : \quad t < t_l \\ \frac{t-t_l+1}{t_r-t_l+1} & : \quad t \in \{t_l = t_0, t_1, \ldots, t_r\} \quad \text{and} \\ 1 & : \quad t > t_r \end{cases} \tag{2.25}$$

$$p(t) = \begin{cases} \frac{1}{t_r-t_l+1} & : \quad t \in \{t_l = t_0, t_1, \ldots, t_r\} \\ 0 & : \quad \text{otherwise.} \end{cases} \tag{2.26}$$

The number of arrivals in an interval when the arrivals occur at a constant rate are often described by the Poisson distribution with

$$A(t) = \frac{e^{-\lambda}\lambda^t}{t!} \quad \text{and} \tag{2.27}$$

$$p(t) = e^{-\lambda} \sum_{i=0}^{t} \frac{\lambda^i}{i!}, \tag{2.28}$$

where $t \in \mathbb{N}$.

Other discrete distributions include the Bernoulli distribution, the binomial distribution, and the negative binomial distribution, among others [107].

## 2.3 Wired Network Architectures

The most important and critical step in designing a communication system for a parallel or distributed system is to choose the network architecture. The architecture is given by the network topology, the buffer sizes, the buffer positions, and so on.

The chosen architecture must fulfill all requirements given by the amount and shape of the expected network traffic, the number of communication partners, and the distance between them. Network architectures are classified as wired or wireless. In this section, it is focused on wired network architectures, while the next section will deal with wireless network architectures. Network architectures for on-chip communication, with their particular characteristics, are introduced in Sect. 2.5.

Many wired network architectures have been proposed [6, 51, 77, 119, 133, 186, 195]. Some of them are briefly discussed below.

### 2.3.1 Basic Classifications

In general, wired networks are categorized as direct or indirect. Direct networks are also called static networks. They consist of a limited number of fixed point-to-point links between some communication partners (nodes). Messages are transferred from a source node (sender) to a destination node (receiver) via intermediate nodes. There are no further switches to change links. For instance, a mesh (Sect. 2.3.3) belongs to this group of networks. The most important characteristic of direct networks is their node degree: the number of links of a node to neighboring nodes.

Indirect networks are also called dynamic networks. They consist of many switches to dynamically change links between nodes. No intermediate nodes are involved in transferring a message from a source node to a destination node. For instance, a crossbar (Sect. 2.3.9) belongs to this group of networks. The most important characteristic of indirect networks is their number of stages: the number of switches of a connection.

The switches in indirect networks may themselves consist of dynamic networks. For instance, smaller crossbars may realize the switches in multistage interconnection networks (Sect. 2.3.10). The nodes of direct networks usually also contain such switches. They connect all inputs and outputs of the node and its core. Thus, direct networks as well as indirect networks may reveal a hierarchical structure.

The way several pairs of nodes communicate in a network without interference is called multiplexing. In wired networks, two kinds of multiplexing are mainly used: multiplexing in space and multiplexing in time.

Space division multiplexing (SDM) separates communication channels of different sender-receiver pairs by space. That means different pairs of nodes use different wires that are not connected to each other. Switches between the wires ensure that nodes of sender-receiver pairs may change if desired. The large quantity of wires needed for a large number of communication pairs is the main disadvantage of SDM.

Time division multiplexing (TDM) allows a sender-receiver pair to use a wire for a certain amount of time. Then, another sender-receiver pair can access this wire for a certain amount of time, and so on. Thus, only a single wire is needed. But TDM suffers from the drawback that all nodes must be synchronized. Only if all nodes deal with exactly the same time, are they able to send or receive in the desired time slice and not interfere with any other node.

Space division multiplexing and time division multiplexing may also be combined. Pure SDM is used in fully connected network architectures where each node is connected to every other node. Pure TDM is used in a single bus architecture.

## 2.3.2 Bus

The most common network architecture to connect few nodes is a bus (Fig. 2.4). Each node is connected to the bus. The source node (called master)

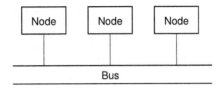

**Fig. 2.4.** Bus architecture

initiates the communication by allocating the bus. It transmits the destination node's address and the message via the bus. All nodes listen to the bus and compare this address with their own. An address match identifies the destination node (called slave), which reads the message. Finally, the bus is deallocated.

Due to the concept of a single common bus, time division multiplexing must be applied and only one sender is allowed to transfer a message at a given time. That is why the bus becomes a bottleneck in a communication system consisting of a large number of nodes.

To overcome this problem, some systems use several busses to combine time division multiplexing and space division multiplexing. Each bus connects only a part of the nodes. All busses are linked to a switching fabric which couples the whole system. Switching fabric architectures are discussed in Sect. 2.3.11.

Busses profit from their simple hardware setup and from their simple routing.

## 2.3.3 Mesh

A popular static network architecture of parallel computers is a mesh [48]. In such an architecture, the nodes are located at the crosspoints of the mesh. Three kinds of meshes are distinguished: one-dimensional meshes (also called chains), two-dimensional meshes (2-D meshes, grids), and three-dimensional meshes (3-D meshes). Figure 2.5 shows a 2-D mesh.

Each node is connected to its two nearest neighbors in each dimension. For instance, four bidirectional links handle all communication of a node of a 2-D mesh. The number of links per node does not change if additional nodes are added to the mesh. Therefore, a mesh offers very good scalability. Additionally, it is of low cost, because a mesh network consists of fewer links per node than most other architectures (crossbars and their hierarchies as an exception are presented later).

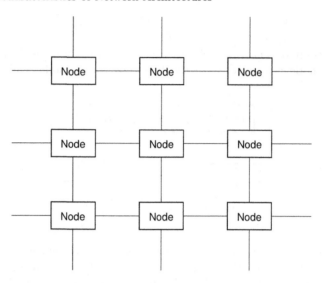

**Fig. 2.5.** 2-D mesh architecture

Their blocking behavior reveals one of the most important disadvantages of meshes. Usually, messages pass several nodes and links before they reach their destination. As a result, links are demanded by many connections using the same link; blocking occurs. Blocking can be reduced if communication is mainly local, e.g., if a task to solve differential equations or finite element methods is spread over nodes. Then, messages are exchanged only between nodes located close together.

Messages are mostly transferred by packet switching, leading to simple routing, another advantage of meshes. For instance, the packet header includes the destination information as $\Delta x$ and $\Delta y$ (in the two-dimensional case) representing the destination node distance in the $x$ direction (horizontal) and the $y$ direction (vertical), respectively. Packets may then be forwarded in the $x$ direction first. The sign of $\Delta x$ determines whether the positive or the negative direction must be chosen. Each intermediate node decrements/increments $\Delta x$. If $\Delta x = 0$ is reached, the packet is forwarded in the $y$ direction in the same manner. $\Delta y = 0$ means that the destination is reached.

This algorithm is called XY routing. Other algorithms are, for instance, the West-First, North-Last, and Negative-First routing, where particular turns are forbidden.

Many variations on the above scheme are known. Forwarding in $x$ direction and $y$ direction may be merged to get alternative paths through the mesh. Partially forwarding in the wrong direction may help avoid blockings or faulty links.

### 2.3.4 Torus

An extension of the mesh architecture is given by the torus. It is a mesh network where all boundary nodes show additional links to their corresponding boundary node at the opposite boundary. Figure 2.6 depicts a two-dimensional torus (2-D torus). Three-dimensional tori (3-D tori) also exist. Higher dimen-

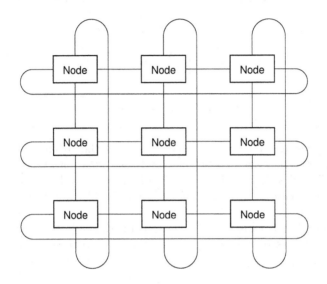

**Fig. 2.6.** 2-D torus architecture

sions are possible, too. The communication network of many parallel computer systems consists of a 3-D torus.

A torus is a static network, like a mesh. Due to their similar structure, a torus reveals the same advantages and drawbacks as a mesh. A slight benefit compared to a mesh is the shorter average distance between two communicating nodes in a torus. The additional links at the boundary avoid long distances between opposite nodes.

### 2.3.5 Ring

Another static network is called ring. In such an architecture, each node is connected to exactly two other nodes, one on each side, leading to an overall structure of a closed loop (Fig. 2.7). Having only two neighboring nodes keeps the amount of interfaces per node very small.

Messages are sent to the ring and usually circle in a common direction from node to node. Each node checks whether it is the receiver. A ring network often realized is the token ring. There, a token synchronizes the nodes' access to the network. The token circles on the ring to signal the status of the network.

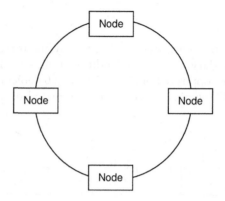

**Fig. 2.7.** Ring architecture

To send a message, nodes must wait to receive the token. If it is marked as "empty," the message can be sent, and the token marking is changed to "occupied." Nodes that receive this token now are not allowed to send. After the message circles around once, it is removed and the token is marked as "empty" again.

The main drawback is that the entire network is affected if any link fails. Doubling each link reduces the problem. Such an architecture is called dual ring.

### 2.3.6 Star

The star network describes a static architecture where all nodes are connected to a central node (Fig. 2.8). Thus, all nodes (except the central one) can only communicate to the others via the central node. That is why the central node may become a bottleneck of a system that consists of many nodes.

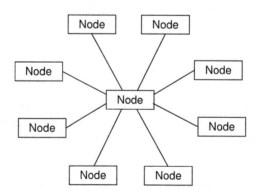

**Fig. 2.8.** Star network architecture

On the other hand, only two hops are needed to reach any of the other nodes. If the central node is sender or receiver, a single hop realizes a communication.

### 2.3.7 Tree

In the static architecture of a tree network, all nodes are arranged as a tree: a root node is connected to descendant nodes. These nodes are again connected to descendant nodes, and so on. Nodes with no further descendant nodes are called leafs.

If the structure of the tree is such that all nodes (except the leafs) are connected to a fixed number $k$ of descendants, the network architecture is called a $k$-ary tree. For instance, Fig. 2.9 shows a binary tree. This tree is

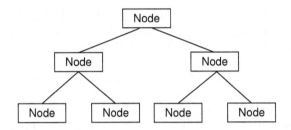

**Fig. 2.9.** Binary tree network architecture

also called balanced because all leaf nodes have the same distance to the root node.

If the nodes of a tree are arranged such that communication mainly takes place between nodes that are located in the same (minimum) subtree, connections consist of only few hops.

On the other hand, if any node in the left half of the tree communicates with any node in the right half, the communication is established via the root node. Therefore, the root node acts as a bottleneck of the network.

An alternative structure, named fat tree [114], overcomes this problem: Nodes are placed only at the leafs of the tree. The nodes at the tree branches are replaced by switches, and the capacity of the connections is increased by $k$ at each stage from the leafs to the root. A dynamic network architecture results. It is topologically equivalent to a bidirectional multistage interconnection network, described in detail in Sect. 2.3.10.

### 2.3.8 Hypercube

A further static network architecture is given by a hypercube [168]. The nodes of the hypercube also represent the nodes of the network. As an example,

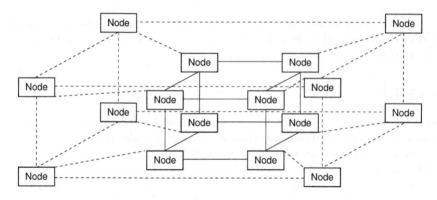

**Fig. 2.10.** 4-D hypercube architecture

Figure 2.10 shows a four-dimensional hypercube (4-D hypercube). It is established by a 3-D hypercube (solid lines) where each node is connected to an additional node of another 3-D hypercube (dashed lines) forming a 4-D hypercube.

Hypercubes can be built of any dimension $d$, connecting $N = 2^d$ nodes. Nevertheless, an existing hypercube cannot be enlarged to a higher dimension unless there are unused ports at the nodes.

Concerning the blocking behavior, equal problems arise as for meshes. But blocking occurs less frequently because messages use less links in average from source node to destination node. At most, $d$ hops are needed to reach a destination.

### 2.3.9 Crossbar

Crossbars [59, 98, 152, 176, 214, 237, 243, 245] are dynamic networks consisting of a switch matrix. This switch matrix ensures that each network input can be connected to each network output via exactly one switch: A connection consists of only a single hop. Figure 2.11 shows a 4×4 crossbar, which consists of four inputs and four outputs. This means that the crossbar connects four nodes. Alternative graphical representations of crossbars are depicted in Fig. 2.12. The right right part of the figure gives the simplest way of representation, and is mainly used.

The switches are located at the crosspoints of the horizontal and vertical lines. Each switch corresponds to a specific input-output pair. The inputs and outputs are connected to the related source nodes (senders) and destination nodes (receivers).

Crossbars of $c$ inputs and $c$ outputs produce $c^2$ switches/crosspoints. This means that the crossbars are scalable, but the complexity increases quadratically; the main drawback of crossbars. Hierarchically connected crossbars, e.g., multistage interconnection networks (Sect. 2.3.10), avoid this drawback.

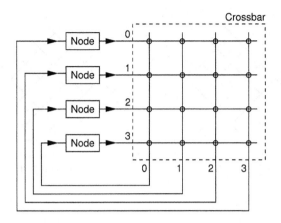

**Fig. 2.11.** Crossbar architecture

If messages are forwarded by packet switching, buffers can be inserted to prevent packet loss in the case of conflicts for resources [30]. Usually, first-in first-out (FIFO) buffers are used. Several buffer locations are appropriate: buffers at crossbar inputs, buffers at crossbar outputs, internal buffers, and shared buffers [250].

Figure 2.13(a) shows a crossbar architecture, including buffers at crossbar inputs (input buffering) [230]. Input buffering suffers from head-of-line (HOL) blocking, for instance, if the first position of all buffers is occupied by packets that are destined to the same crossbar output. Then, all packets except one are blocked, and most parts of the crossbar remain idle. Even if the second buffer position contains a packet destined to another output, the crossbar mainly remains idle because this packet cannot pass the blocked packet.

Buffers at the outputs (output buffering) [123, 249] overcome this problem (Fig. 2.13(b)). The major drawback of such a buffer location is the high packet transfer rate of the crossbar that is needed, for instance, if all $c$ crossbar inputs

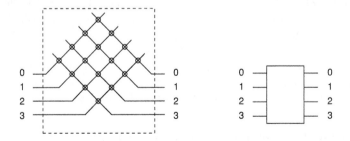

**Fig. 2.12.** Alternative crossbar representations

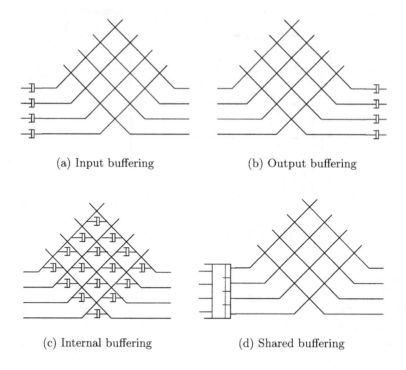

(a) Input buffering                  (b) Output buffering

(c) Internal buffering               (d) Shared buffering

**Fig. 2.13.** Crossbar buffer locations

receive a packet destined to the same output. Then, the transfer rate of the crossbar must be $c$ times as high as the input rate at the crossbar.

Internal buffers overcome both previously mentioned problems. Figure 2.13(c) depicts such a buffer scheme. But there are also drawbacks to this architecture: a much higher number of FIFO buffers ($c^2$) is needed. This means that hardware cost is increased.

A completely different buffering scheme is given by shared buffers [177, 210]. Shared buffers can be located at crossbar inputs (Fig. 2.13(d)), crossbar outputs, or internally. A common pool of buffer space is available. Each input/output allocates buffer space as required.

The drawback of such a buffering scheme emerges from the higher control cost: the buffer must manage where to store which packet. Internal queues have to be used to keep the transfer order of the packets.

### 2.3.10 Multistage Interconnection Network

To overcome the drawbacks of crossbars, multistage interconnection networks (MIN) [1, 55, 71, 87, 99, 104, 119, 140, 179, 235, 242] are frequently proposed to connect the nodes of parallel systems or to establish switching fabrics (Sect. 2.3.11) connecting the nodes of distributed systems.

Such architectures were already used to design telephone switches in the 1960s. At that time, Beneš [13, 14] investigated MINs operating in circuit switching mode and established a mathematical description of their behavior. Nowadays, buffers are inserted [250] and packet switching is applied to MIN architectures that are part of computer networks.

MINs are dynamic networks based on switching elements (SEs) [84]. The most common approach to realize SEs are crossbars. SEs are arranged in stages and connected by interstage links. The link structure and number of SEs characterizes the MIN. Several MIN architectures exist.

## MIN with Banyan Property

Multistage interconnection networks with the banyan property [33, 49, 85, 103, 179] are networks where a unique path from an input to an output exists. Such MINs of size $N \times N$ ($N$ inputs and $N$ outputs) consist of $c \times c$ switching elements (SEs of $c$ inputs and $c$ outputs) [32, 156, 216] with $n = \log_c N$ stages (Fig. 2.14).

To achieve synchronously operating switches, the network is internally clocked [246]. This network clock cycle consists of as many hardware clock cycles as are needed to completely forward all phits of a packet for one stage.

At each stage $k$ ($0 \leq k \leq n-1$), there is a FIFO buffer of size $m_{max}(k)$ in front of each SE input [50, 142, 206, 244]. Of course, output buffering, internal buffering, or shared buffering are also possible. The packets are forwarded by store-and-forward switching, cut-through switching, or wormhole switching from one stage to its succeeding one.

Packets that are destined to full buffers can be handled by dropping those packets [240] or by applying the backpressure mechanism. The backpressure mechanism keeps packets in their current stage until the required buffer becomes available again. This means that no packets are lost within the network. Local and global backpressure are distinguished. Local backpressure observes only the destination buffer at the next stage: the packet at stage $k$ is sent if space at stage $k + 1$ is available. Global backpressure acquires additional information about packet flows: the packet at stage $k$ is sent even if no space at stage $k + 1$ is available, but will become available by the time the packet is received. Such a situation may arise if a packet leaves stage $k + 1$ at the same clock cycle.

The network shown also belongs to the class of delta networks. This means that it is a banyan network where all packets can use the same routing tag to reach a certain network output independently of the input at which they enter the network. Rectangle delta networks additionally demand square SEs (i.e., equal number of SE inputs and outputs), shown in Fig. 2.14. This network also belongs to the class of regular delta networks (i.e., equal size of all SEs)

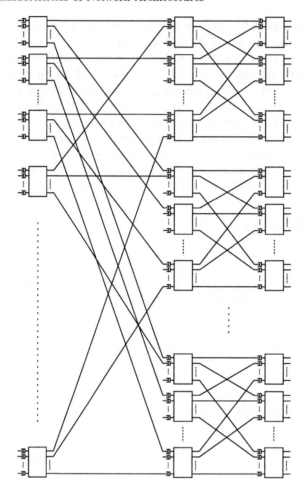

**Fig. 2.14.** Three-stage delta network consisting of $c \times c$ SEs

and bidelta networks (i.e., delta property for input-to-output direction and output-to-input direction).

Various bidelta network architectures consisting of $2 \times 2$ SEs are depicted in Fig. 2.15. In the literature [102, 103], they are referred to as Omega, Flip, Baseline, Indirect Binary Cube (IBC), and Modified Data Manipulator (MDM). Their interstage connections distinguish them. For instance, the interstage connections of the Modified Data Manipulator are established by numbering the SEs at each stage $k$. Numbers are coded to the base of $c$, starting with 0. This means that each SE is numbered by a $(n-1)$-digit number $\nu_{n-1}\nu_{n-2}\ldots\nu_2\nu_1$, where $0 \leq \nu_i \leq c-1$ and $1 \leq i \leq n-1$. Then, SE $\nu_{n-1}\nu_{n-2}\ldots\nu_2\nu_1$ at stage $k$ is connected to $c$ SEs at stage $k+1$ that are numbered $\nu_{n-1}\ldots\nu_{n-k} \diamond \nu_{n-k-2}\ldots\nu_1$, where $\diamond$ equals all values from 0 to

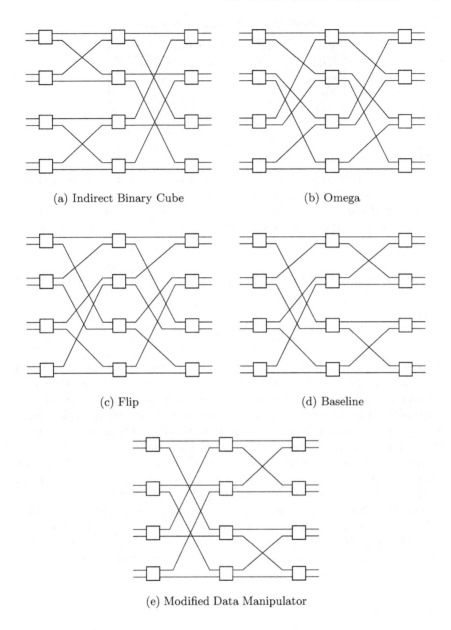

(a) Indirect Binary Cube    (b) Omega

(c) Flip    (d) Baseline

(e) Modified Data Manipulator

**Fig. 2.15.** Bidelta networks consisting of 2×2 SEs

$c - 1$. The rules concerning the interstage connections of all architectures of Fig. 2.15 are given in Table 2.2.

**Table 2.2.** Interstage connections

| Bidelta network | SEs at stage $k+1$ |
|---|---|
| MDM | $\nu_{n-1} \ldots \nu_{n-k} \diamond \nu_{n-k-2} \ldots \nu_1$ |
| Omega | $\nu_{n-2} \ldots \nu_1 \diamond$ |
| Flip | $\diamond \nu_{n-1} \ldots \nu_2$ |
| Baseline | $\nu_{n-1} \ldots \nu_{n-k} \diamond \nu_{n-k-1} \ldots \nu_2$ |
| IBC | $\nu_{n-1} \ldots \nu_{k+2} \diamond \nu_k \ldots \nu_1$ |

Alternatively, mathematical formulae can be applied to obtain the interstage connections. Then, all SE inputs and all SE outputs at every stage must be numbered consecutively from 0 to $N-1$. For instance, in MDM architecture, the output $\ell_O(k)$ at stage $k$ is connected to input $\ell_I(k+1)$ at stage $k+1$ by

$$\ell_I(k+1) = \ell_O(k) - \left( \left( \zeta \text{ div } \frac{N}{c^{k+1}} \right) - (\zeta \text{ mod } c) \right) \cdot \left( \frac{N}{c^{k+1}} - 1 \right), \quad (2.29)$$

where

$$\zeta = \ell_O(k) - \frac{N}{c^k} \cdot \left( \ell_O(k) \text{ div } \frac{N}{c^k} \right). \quad (2.30)$$

Besides the ways presented on how interstage connections are established, many other architectural variations exist. They are mainly developed to reduce the amount of blocked packets due to occupied resources like buffers or SEs. Some of them are presented below. All of them offer multiple paths between a source-destination pair: if a packet will be blocked due to an occupied resource, it can choose an alternative path to the destination.

Besides avoiding blocking, alternative paths also help deal with faulty network elements. If a path of a packet crosses a faulty network element, it can be changed to an alternative path to the packet's destination.

On the other hand, abandoning the banyan property may cause out of order packet sequences: packets of a message are not received in the order sent. They may pass others by taking alternative paths through the network. Then, the destination node must be able to deal with such a scenario, e.g. by ordering the packets again.

### Dilated MIN

Dilation reduces blocking by replicating the interstage connection lines $d$ times. Then, the SE size must be increased by a factor of $d$ to ensure the required number of SE inputs and outputs for the connection lines: $c \times c$ SEs of a MIN with banyan property result in $(c \cdot d) \times (c \cdot d)$ SEs for the dilated MIN. Figure 2.16 shows the architecture of an $8 \times 8$ dilated MIN with all interstage connection lines doubled ($d = 2$). The concept of dilated MINs was

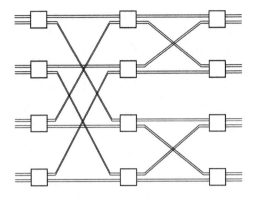

**Fig. 2.16.** Dilated multistage interconnection network ($d = 2$)

introduced by Kruskal and Snir [102]. It allows transmitting up to $d$ packets from a particular SE at stage $i$ to any SE at stage $i + 1$. Blocking occurs only if more than $d$ packets are sent or if the destination buffer does not provide sufficient space.

At each network output, a $d$:1 multiplexer collects the packets from the corresponding SE outputs at the last network stage and forwards them to the network output. Two different output schemes are distinguished: single acceptance (SA) and multiple acceptance (MA). Single acceptance means that just one packet is accepted by the network output per clock cycle. If there are packets at more than one corresponding SE output, one of them is chosen. All others are blocked at the last stage. The multiplexer decides according to its scheduling algorithm which packet to choose.

Multiple acceptance means that more than one packet may be accepted by the network output per clock cycle. Either all packets are accepted or just an upper limit $R$. If an upper limit is given, $R$ packets are chosen to be forwarded to the network output and all others are blocked at the last stage. As a result, single acceptance is a special case of multiple acceptance with $R = 1$.

### Replicated MIN

Replicated MINs enlarge multistage interconnection networks with the banyan property by replicating them $L$ times. The resulting MINs are arranged in $L$ layers. Corresponding input ports are connected, as well as corresponding output ports. Figure 2.17 shows the architecture of an $8 \times 8$ replicated MIN consisting of $2 \times 2$ SEs and two layers. A three-dimensional view of the same network is given by Fig. 2.18. As with dilated MINs, replicated ones were also introduced by Kruskal and Snir [102]. Packets are received by the inputs of the network and distributed to the layers. Layers may be chosen by random, by round robin, dependent on layer loads, or any other scheduling algorithm. The distribution is performed by a 1:$L$ demultiplexer.

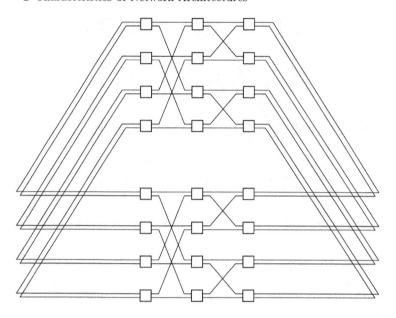

**Fig. 2.17.** Replicated multistage interconnection network ($L = 2$)

At each network output, an $L{:}1$ multiplexer collects the packets from the corresponding layer outputs and forwards them to the network output. Two different output schemes are distinguished, as in the case of dilated MINs: single acceptance (SA) and multiple acceptance (MA). If single acceptance is applied and there are packets in more than one corresponding layer output, one of them is chosen. All others are blocked at the last stage of their layer. The multiplexer decides according to its scheduling algorithm which packet to choose.

Multiple acceptance also works similar to this scheme in case of dilation: either all packets are accepted, or just an upper limit $R$. If an upper limit is

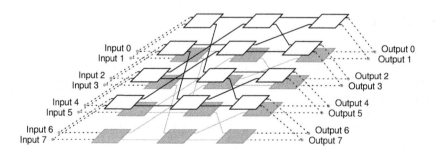

**Fig. 2.18.** Replicated multistage interconnection network ($L = 2$, 3-D view)

given, $R$ packets are chosen to be forwarded to the network output, and all others are blocked at the last stage of their layer. As a result, single acceptance is a special case of multiple acceptance with $R = 1$.

Replicated MINs may avoid out of order packet sequences by sending packets belonging to the same connection to the same layer.

### Beneš Network

Another approach to reduce blocking is to add further network stages. Figure 2.19 shows the Beneš network [13] as an example. It is of size $8 \times 8$ and consists

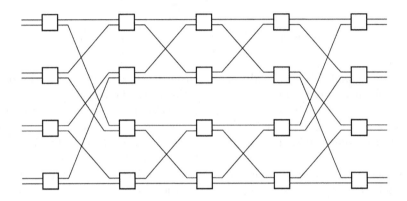

**Fig. 2.19.** Beneš network

of $2 \times 2$ SEs. A Beneš network is built by extending a Baseline network with its inverse one. The last stage of the Baseline and the first stage of the inverse Baseline are merged. This leads to $2n - 1$ stages for Beneš networks. Instead of a Baseline, all other MIN architectures revealing the delta property can also serve as a network basis.

Due to their architecture, Beneš networks are non-blocking multistage interconnection networks if packet switching is performed. They show the smallest complexity of all non-blocking packet-switched MINs.

### Bidirectional MIN

Bidirectional MINs [1, 239] operate with SEs and interstage connection lines in which packets can pass in both directions. As a result, such networks offer alternative paths for all source-destination pairs. Figure 2.20 shows an $8 \times 8$ bidirectional MIN consisting of $2 \times 2$ bidirectional SEs. In contrast to previously presented MINs, sources and destinations are located at the left-hand side of the figure, due to the bidirectional connections.

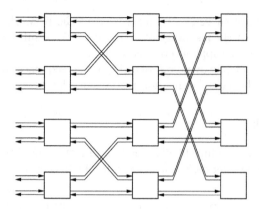

**Fig. 2.20.** Bidirectional multistage interconnection network

Bidirectional SEs need additional transfer features compared to ordinary SEs which had only to transfer packets in the forward direction. Bidirectional SEs must also be able to transfer in the backward direction and to allow turnarounds (Fig. 2.21). Due to those turnarounds, bidirectional MINs are also called turnaround MINs. If their SEs are crossbars, they are called turnaround crossbars.

Depending on their source and destination nodes, packets pass between one and $2n - 1$ stages to reach their destination.

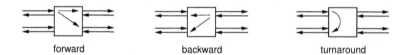

forward              backward              turnaround

**Fig. 2.21.** Transfer directions of bidirectional switching elements

## Tandem Banyan Network

Tandem banyan networks [180, 197] are established by placing multiple MINs with the banyan property in sequence (Fig. 2.22). Each output of the individual MINs is connected to the corresponding input of the following MIN. Additionally, it is connected to the corresponding overall output of the tandem banyan network (Fig. 2.22, bottom).

According to the banyan property, packets try to follow their unique path through the first MIN. If a packet reaches an SE where further packets are destined to the same SE output and it loses the competition for this output, no blocking occurs, in contrast to an ordinary MIN. In the case of a tandem banyan network, the packet is marked and sent to the wrong SE output.

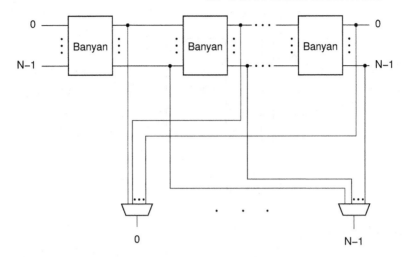

**Fig. 2.22.** Tandem banyan network

Sending to the wrong output is called deflection routing. Then, all packets continue through the MIN. If further conflicts occur, marked packets have minor priority compared to unmarked.

Marked packets that reach the output of an individual MIN are known to leave the MIN at the wrong output. Therefore, they try again to reach the right output by entering the following MIN. Their marking is removed. If they again lose a conflict and reach a wrong output, they enter the following MIN, and so on.

Unmarked packets that reach an output of an individual MIN have never taken the wrong way in this MIN. This means that they have reached the destination output: they are sent to the overall output of the tandem banyan network.

**Recirculation Network**

Recirculation networks [192] are also based on the concept of deflection routing. The way packets that take the wrong way are handled is almost identical to that of tandem banyan networks. But in contrast to tandem banyan networks, recirculation networks consist of only a single MIN with the banyan property. Packets that leave this MIN at the wrong output are recirculated and enter this MIN again (Fig. 2.23). As a result, the hardware costs are heavily reduced.

On the other hand, such architecture shows bad performance in the case of heavy network traffic (Sect. 2.2). Recircled packets superpose with new packets to a large extent and cause network congestion.

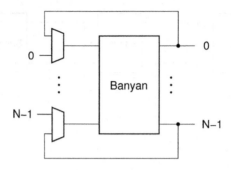

**Fig. 2.23.** Recirculation network

## Sorter Network

Blocking may also be reduced by sorter networks [145]. Sorter networks are inserted in front of a MIN with banyan property. Packets entering the sorter network are ordered according to their destination node. As a result, the following MIN with banyan property becomes non-blocking for those packets.

Figure 2.24 shows an 8×8 Batcher-Banyan network as an example. The binary comparators of the Batcher sorter ensure the sorting of the packets before they enter the MIN with banyan property. Inactive comparator inputs operate as if covering a packet that is destined to an output number larger than the highest available output number.

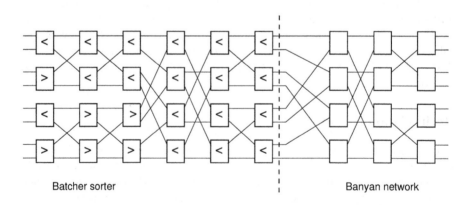

**Fig. 2.24.** Batcher-Banyan network

**Expanded Delta Network**

To overcome blocking in delta networks, expanded delta networks [7] are introduced. Such networks are established by interleaving several networks with the delta property. The quantity of interleaved delta networks is called expansion factor *EF*. It also describes the number of output lines per output. For example, a 2×2 expanded delta network with $EF = 8$ is given in Fig. 2.25. Such a

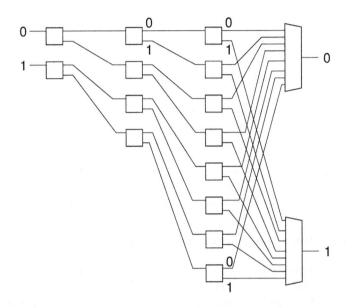

**Fig. 2.25.** Expanded delta network

network is also known as Expanded Delta Fast Packet Switch (EDFPS).

The main drawback of those networks are the huge hardware costs. Even the costs of a 2×2 network are high, as can be seen from Fig. 2.25.

**Clos Networks**

Clos networks [37, 67, 184, 200, 219] are characterized by how switches at the middle stage are connected to the first and last network stages. The basic version of a Clos network consists of three stages, as shown in Fig. 2.26. An $N \times N$ network where $N = k \cdot s$ is built of $k$ SEs of size $s \times m$ at the first stage, $m$ SEs of size $k \times k$ at the middle stage, and $k$ SEs of size $m \times s$ at the last stage. Each network input-output pair can be connected by a path via an arbitrary middle stage SE. This means that $m$ paths are available for each connection. A non-blocking Clos network is achieved by $m > 2s - 1$.

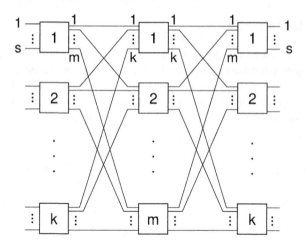

**Fig. 2.26.** Clos network

Clos networks of more than three stages emerge by substituting again the middle stage SEs by a Clos network. Using this scheme multiple times, Clos networks with an arbitrary but odd number of stages can be established.

### Multilayer Multistage Interconnection Networks

The architecture of multilayer multistage interconnection networks (MLMIN) [218] resulted from the idea to create an optimal MIN that is able to deal with multicast traffic. Multicast traffic evokes a high network load because each multicast message is delivered to many destination nodes (see Sect. 2.2.1). This means that each multicast message is multiplied before or during its journey through the MIN.

Replicated networks as presented in Sect. 2.3.10 could be a solution: replicated MINs of $L$ layers handle traffic with an average of $L$ multicast destinations with similar performance as that of regular MINs with the banyan property that transfer unicast traffic. But such networks show $L$ times higher cost.

A better solution is to profit from the particular way in which a multicast is performed in MINs if packet replication while routing is applied. Figure 2.27 shows such a scenario for an 8×8 MIN consisting of 2×2 SEs. A packet is received by Input 3 and destined to Output 5 and Output 7. The packet enters the network and is not copied until it reaches the middle stage. Then, two copies of the packet proceed through the remaining stages.

Packet replication before routing in the above example would copy the packet and send it twice into the network. Therefore, packet replication while routing reduces the number of packets at the first stages. In other words, comparing the packet density at the stages in case of replication while routing

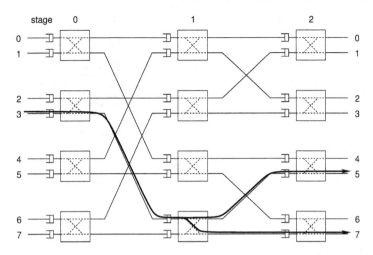

**Fig. 2.27.** Multicast while routing

shows that the greater the stage number, the greater is the number of packets. The only exception is when the traffic pattern results in such a destination distribution that packet replication always takes place at the first stage. Then, the number of packets is equal at all stages. But such a distribution is very unlikely, in general.

To set up multistage interconnection networks that are appropriate for multicasting, the previously mentioned different traffic densities of the stages must be considered.

The newly developed multilayer multistage interconnection networks (MLMINs) take care of these multicast traffic characteristics. Due to the different traffic densities, more switching power is needed at the last stages, compared to the first stages, of a network.

To supply the network with the required switching power, the new network structure replicates the number of layers at each stage. The factor with which the number of layers is increased is called growth factor $G_F$ ($G_F \in \mathbb{N}\backslash\{0\}$). Figure 2.28 shows an 8×8 MLMIN (three stages), with growth factor $G_F = 2$, in lateral view. That means the number of layers is doubled at each stage, and each switching element has twice as many outputs as inputs. Assume, for example, that 2×2 SEs are used. Such an architecture ensures that even in case of two broadcast packets staying at the inputs, all packets can be sent to the outputs (if there is buffer space available at the succeeding stage). For instance, one packet is broadcast to the lower layer and the other one to the upper layer. Figure 2.29 shows such a scenario. As can be seen in the figure, the SE architecture of MLMINs slightly differs from regular SEs: the outputs are replicated by the growth factor $G_F$ to connect the switch to all $G_F$ layers.

Replicating the number of layers at each stage avoids unnecessary layer replications at the first stages, as in case of the replicated MINs. Choosing

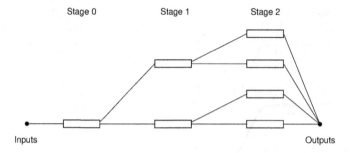

**Fig. 2.28.** Multilayer multistage interconnection network ($G_F = 2$)

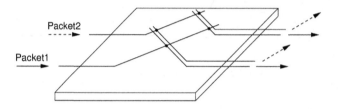

**Fig. 2.29.** Unblocked broadcast at a 2×2 SE with $G_F = 2$

$G_F = c$ ensures that no internal blocking occurs in an SE, even if all SE inputs broadcast their packets to all SE outputs. Nevertheless, blocking may still occur at the network output, depending on the number $R$ of accepted packets per clock cycle.

A drawback of the new architecture arises from the exponentially growing number of layers for each additional stage. The more the number of network inputs established, the more the stages and layers. To limit the number of layers and, therefore, the amount of hardware, two options are considered: starting the replication in a more rear stage and stopping further layer replication if a given number of layers is reached.

The first option is demonstrated in Fig. 2.30 in lateral view. The example presents an 8×8 MLMIN in which replication does not start before Stage 2 (last stage), with $G_F = 2$. A 3-D view is given in Fig. 2.31. The stage number

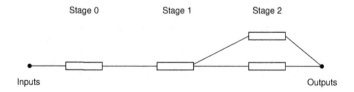

**Fig. 2.30.** MLMIN in which replication starts at Stage 2 (lateral view)

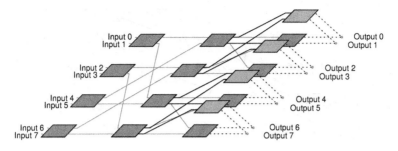

**Fig. 2.31.** MLMIN in which replication starts at Stage 2 (3-D view)

in which replication starts is defined by $G_S$ ($G_S \in \mathbb{N}$). Figures 2.30 and 2.31 introduce an MLMIN with $G_S = 2$. Of course, moving the start of layer replications some stages to the rear not only reduces the number of layers, it also reduces the network performance due to fewer SEs and, therefore, fewer paths through the network [218].

Stopping further layer replication if a given number $G_L$ of layers is reached also reduces the network complexity ($G_L \in \mathbb{N}\backslash\{0\}$). It prevents exponential growth beyond reasonable limits in the case of large networks. Figure 2.32 shows such an MLMIN, with limited number of layers in lateral view. A 3-D view is presented in Fig. 2.33. The number of layers of this 8×8 MLMIN is

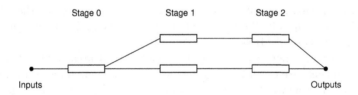

**Fig. 2.32.** MLMIN with limited number of layers (lateral view)

limited to an upper $G_L = 2$. Layers are replicated with a growth factor of $G_F = 2$. As in the previous option, the reduced amount of SEs also decreases network performance [218].

Both presented options can be combined to further reduce network complexity. Such a network is determined by parameters $G_S$ (start of replication), $G_F$ (growth factor), and $G_L$ (layer limit). For instance, Fig. 2.33 shows an MLMIN with $G_S = 1$, $G_F = 2$, and $G_L = 2$.

MINs with the banyan property and replicated MINs can be considered as special cases of MLMINs. MINs with the banyan property are equivalent to MLMINs with $G_F = 1$. In this case, $G_S$ and $G_L$ have no effect. Replicated MINs are equivalent to MLMINs with $G_S = 0$, $G_F = L$, and $G_L = L$.

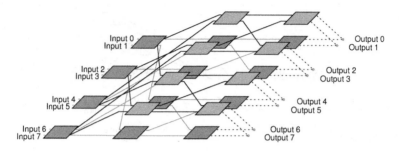

**Fig. 2.33.** MLMIN with limited number of layers (3-D view)

An exhaustive comparison of MLMIN cost and MLMIN performance to MINs with the banyan property and replicated MINs can be found in [218]. There, the advantage of the performance related to the cost of multilayer multistage interconnection networks is elaborated.

### 2.3.11 Switching Fabric

Often, network nodes or subnetworks are connected by a central device. Such a device is called switch [6]. A switching fabric often builds the internal structure of a switch.

Most switching fabrics are based on multistage interconnection networks. All dynamic network architectures of Sect. 2.3.10 that overcome the problem of blocking are suitable. These include crossbars (Sect. 2.3.9), which can be seen as special case of MINs: a one-stage MIN.

Many irregular network structures, such as local area networks (LANs) and the Internet, are established using switches and switching fabrics. Nodes are connected by a switch building a subnetwork. Subnetworks are again connected by a switch building a larger subnetwork. These larger subnetworks are again connected by a switch, and so on.

### 2.3.12 Dynamic Networks versus Static Networks

Previously presented network architectures were classified as dynamic (indirect) and static (direct) (Sect. 2.3.1). Static networks were characterized as having a limited number of fixed point-to-point links between nodes while dynamic networks consist of many switches to dynamically change links between nodes.

But looking at the nodes of static networks in more detail reveals that there are also switches to change connections. For instance, Fig. 2.34 shows a 2-D mesh and one of its nodes. The nodes of the mesh incorporate a node core and a 5×5 switching element (Fig. 2.34(b)), optionally with buffers. The switching element connects all inputs and outputs of the node to allow packets that are

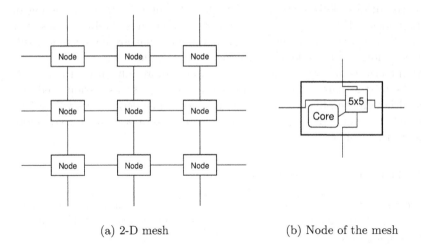

(a) 2-D mesh                    (b) Node of the mesh

**Fig. 2.34.**  2-D mesh architecture

sent from a neighboring node to pass the node in question. Furthermore, the node core, which represents the actual node functionality, is linked via the switching element to the rest of the mesh.

The switching element can be realized by a crossbar, for instance. Thus, one should be aware that the classification in dynamic networks and static ones is not clearly distinct.

## 2.4 Wireless Network Architectures

Besides wired network architectures, as presented in the previous section, wireless architectures [25, 174, 190] may also be used to establish a network for a distributed system. This means that the nodes of a network are not connected by wires. Information is exchanged via the transmission of electromagnetic waves. The detailed physical background about electromagnetic waves will not be discussed here. The interested reader is referred to [150, 185]. This section addresses the architectures of wireless networks and their dynamic reconfiguration due to moving nodes.

Often, network architectures are characterized as fixed or mobile [174]. These attributes refer to the nodes of the network. Fixed nodes do not change their location. Thus, they are usually connected to a network by wire. Nevertheless, there also exist fixed nodes with wireless connections, for instance, because a wired connection would be more expensive than a wireless one.

In contrast to fixed nodes, mobile nodes move. This leads to changing network architectures, because different locations may lead to different network access conditions. This section deals with distributed systems consisting of mobile nodes (MNs) and wireless network architectures. Mobile nodes that have a wired network connection (e.g., notebooks that are connected at different locations via a modem or Ethernet) are not addressed.

Architectures of wireless networks are much more influenced than wired networks by the particular constraints given by the transmission medium. Thus, although the physical background of electromagnetic waves will not be discussed here, some basics about wireless transmission are introduced before the network architectures are presented.

### 2.4.1 Basics in Wireless Transmission

In wireless networks, communication is performed by sending and receiving signals realized by electromagnetic waves. This section discusses the basics of signal propagation.

### Spatial Transmission and Disturbances

To transmit signals as electromagnetic waves, antennas are needed. They build the interface between the communication device and the transmission medium (air, space). The simplest (theoretical) antenna is a point, symmetrically sending signals in all directions with equal power. Such an antenna is called an isotropic radiator.

A real antenna is for instance the Hertzian dipole, a thin metal baton. A length of half of the transmission wavelength $\lambda$ is very effective. The dipole shows uniform (omni-directional) sending power only in the plane perpendicular to the baton.

More complex antennas transmit in preferred directions. This supports the adaption of the transmission to an asymmetric environment, which usually occurs in real-world scenarios. Such antennas are called directional antennas.

By combining multiple antennas, antenna architectures with various characteristics can be realized. These architectures, such as sectorized antennas and multi-element antenna arrays are explained in [174], along with smart antennas that use signal processing to improve performance.

If uniform transmission in all directions in a plane around the sender is assumed, the plane can be divided into three parts (Fig. 2.35): transmission range, detection range, and interference range.

The transmission range describes the area where a receiver can communicate well with the sender due to a good signal noise ratio. The power of the received signal gets weaker the farther the receiver is from the sender. If no communication is possible due to the large distance from the sender, but the sender's signal can still be distinguished from background noise, the area is called detection range. Finally, if the receiver is located within the interference range, the sender can no longer be distinguished from the background noise.

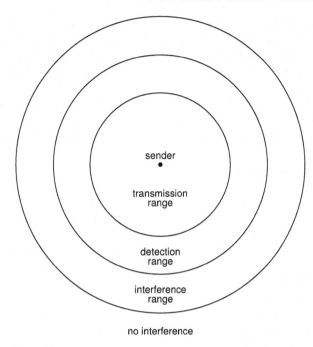

sender
•

transmission
range

detection
range

interference
range

no interference

**Fig. 2.35.** Ranges of a sender

Nevertheless, it is still strong enough to interfere with other signals and to disturb them.

Even in the case of a uniform transmission, the ranges usually do not look like rings, as in Fig. 2.35. That is because objects in this area disturb the signal transmission. The higher the frequency of the signal, the more s the signal disturbed by an object (called obstacle) that is located in the line of sight. The line of sight (LOS) is defined by a straight line between the sender and the receiver. A receiver that is located behind an object does not receive the signal. It is completely blocked by the object (this is called shadowing).

Another disturbance is called reflection. Huge objects reflect the signal while absorbing a part of the power. The reflected signal may interfere with the original signal. Because both signals took paths of different lengths, they arrive at the receiver at different times. Such an effect is called delay spread. In extreme cases, the signal may be destroyed due to interference, particularly if many reflected signals overlay. On the other hand, reflection is often used to reach receivers that are not in the LOS. The signal may be reflected several times to be forwarded to a receiver.

Refraction may also disturb the signal. If the transmission medium changes, refraction occurs due to the different velocities of the electromagnetic wave, depending on the medium.

In the case of objects that are in their size close to the wavelength, scattering and diffraction may occur. Scattering means that the signal is divided into many weaker signals. Diffraction arises if the signal is deflected at an edge of an object. In both cases, the signal power is weakened and strongly depends on the location.

The signal is also influenced if either sender or receiver moves. Then, effects like the Doppler shift must be considered. The Doppler shift describes the wavelength shift between sender and receiver caused by different velocities.

### Multiplexing and Modulation

Wireless communication must deal with two main shortcomings concerning the transfer of information. First, only a single medium is available for all nodes that like to communicate. This problem is solved by multiplexing. Second, information (like '0' and '1' bits) cannot directly be transmitted because electromagnetic waves of particular frequencies must be used. Modulation overcomes this. Both issues are briefly addressed here.

Multiplexing ensures that multiple nodes can use the same medium for communication. Four kinds of multiplexing can be applied to wireless networks: multiplexing in space, in frequency, in time, and in code.

Space division multiplexing (SDM) profits from the fact that the signal power decreases the greater the distance to the sender (Fig. 2.35). This means that if the spatial distance between two senders exceeds a particular limit (twice the interference range), they will not interfere with each other. A node receives the information of the sender in whose range it is located. Communication channels of different sender-receiver pairs are divided by space.

Frequency division multiplexing (FDM) divides the available frequency band into multiple non-overlapping subbands. Thus, communication channels of different sender-receiver pairs are divided by frequency. Guard spaces (very small frequency bands) between adjacent subbands prevent channel interference. To ensure a particular transmission quality and rate, the frequency band can only be split into a small number of subbands. Therefore, only a small number of communication pairs are possible.

Time division multiplexing (TDM) allows a sender-receiver pair to use the entire frequency band for a certain amount of time. Then, another sender-receiver pair can access the frequency band for a certain amount of time, and so on. Gaps between succeeding transmissions serve as guard spaces to prevent interference. TDM suffers from the drawback that all nodes must be synchronized. Only if all nodes are synchronized in time are they able to send or receive in the desired time slice and not interfere with any other node.

Code division multiplexing (CDM) does not divide frequency band or time. All sender-receiver pairs use the same frequency band all the time. Different communication channels are distinguished by codes. Each channel is encoded by a particular code. A receiver only knows the code assigned to its channel. Therefore, it is able to decode the information only of its channel. The other channels form a kind of background noise. In CDM, sender and receiver need some computing power to encode and decode the information.

After one of the above medium access methods is chosen, modulation must be applied to include the information to be transmitted into the electromagnetic waves. To transmit digital data, two steps must be performed. First, a digital modulation transfers the digital data into an analog baseband signal. Then, an analog modulation shifts this signal to the frequency band that is intended for transmission. The digital and analog modulation schemes are briefly presented in the following. A detailed description can be found in [76, 150, 174, 185, 229].

Amplitude shift keying (ASK) is a digital modulation scheme in which the two values '0' and '1' are represented by two different amplitudes of a sine signal.

Another digital modulation scheme is called frequency shift keying (FSK). For instance, in binary FSK (BFSK), the two different binary values result in two different frequencies of the signal. To avoid fast phase changes due to the different frequencies, continuous phase modulation (CPM) and (particularly) minimum shift keying (MSK) are used. This scheme also considers the previous binary value, besides the current one, to determine the frequency.

Phase shift keying (PSK) changes the phase of the signal depending on the digital value. In case of binary PSK (BPSK), the phase is shifted by $\pi$ each time the binary value changes from '0' to '1', and vice versa. Quadrature PSK (QPSK) represents two bits by a particular phase. As a result, four different phases are applied. Previous kinds of PSK schemes need synchronization between sender and receiver to get a common reference signal that can be compared with the current phase of a signal. Differential QPSK (DQPSK) shifts the phase relative to the phase of the previous two bits. No reference signal is needed. Combining PSK with ASK leads to quadrature amplitude modulation (QAM).

Another concept of digital modulation is called multi-carrier modulation (MCM). A bit stream is divided into many bit streams of lower rate. Each of these lower rate bit streams is modulated by one of the previously mentioned schemes, e.g., QPSK, and transmitted via a separate frequency. Orthogonal frequency division multiplexing (OFDM) chooses the frequencies of MCM such that they are orthogonal. This means that the amplitude of each signal reaches its maximum at a frequency at which all other amplitudes are zero. Adding some coding for error detection and correction of the other bit streams results in coded OFDM (COFDM).

After performing the digital modulation, an analog modulation shifts the signal to the desired frequency band. Three analog modulation schemes are mainly used: amplitude, frequency, and phase modulation. They operate similarly to the digital modulation schemes ASK, FSK, and PSK, but in a continuous way. Amplitude modulation (AM) changes the amplitude of the carrier sine signal depending on the amplitude of the analog signal to be coded. Frequency modulation (FM) changes the frequency of the carrier signal to encode the amplitude of the analog signal, and phase modulation (PM) changes the phase of the carrier signal.

This digital and analog modulation result in a radio signal that a sender node of a distributed system transmits to a receiver node. The receiver node transforms the signal via analog demodulation to an analog baseband signal. Depending on the digital modulation of the data, synchronization is needed, for example, to detect a phase shift. With this information, the receiver can recover the original data from the baseband signal.

Often, the bandwidth needed for the radio signal is enlarged before transmission. This technique is called spread spectrum. It helps reduce the interference of signals of low bandwidth called narrowband interference. The two realizations direct sequence spread spectrum (DSSS) and frequency hopping spread spectrum (FHSS), which change the coding of the digital data or change the frequency of the radio signal, respectively, are explained in [159, 174].

### 2.4.2 Cellular Networks

Wired network architectures of parallel and distributed systems are distinguished by considering the location of their nodes and the way how they are connected by wires. Wireless networks use only one medium. The locations of the nodes vary if mobile nodes (MNs) are involved. Thus, wireless network architectures must be classified in a different way than wired ones. This book considers two basic architectures: architectures in which mobile nodes communicate only directly with stationary network elements and architectures that consist of only mobile network elements. The latter architecture will be discussed in the next section.

Architectures in which mobile nodes directly communicate with only the stationary network elements are, for instance, cellular networks [12, 131, 174, 223]. The stationary elements are called base stations (BSs). Each of them covers a particular area (transmission range) by its transmissions. This area is denoted as cell. A mobile node within this cell communicates only with the base station directly. If the mobile node needs to communicate with some other mobile node in the cell, communication is established via the base station.

To cover an area that is larger than the transmission range of a base station, multiple base stations are positioned so that each location of the requested area can be reached by at least one base station. If it is assumed that the transmission range represents a circle, as in Fig. 2.35, the range of each base station can be approximated by a hexagon. Then, the optimal spatial distribution of the base stations is as depicted in Fig. 2.36. Nevertheless, assuming the transmission range to be circles does not always correspond to reality (see Sect. 2.4.1).

Cell sizes are usually kept small to achieve high capacity: cellular networks use space division multiplexing. The greater the number of cells into which the area is divided, the greater the number of communication channels that can

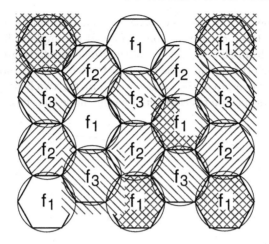

**Fig. 2.36.** Spatial distribution of base stations

simultaneously be established by reusing the same frequency. If the expected density of mobile nodes is small in any area, the cell size may be chosen larger than that in an area where the expected density of mobile nodes is high.

Small cell sizes also lead to lower transmission power needed by the mobile nodes: they are always close to a base station. Because mobile nodes cannot permanently be connected to a power supply network due to their mobility, saving power is an important issue for them.

Another argument for small cell sizes is interference. In small cells, only local disturbances interfere with the transmitted signal. Furthermore, if a cell fails, only a small area is not covered by a base station. Mobile nodes can reach a covered area fast.

On the other hand, small cell sizes cause high cost in infrastructure. Many more base stations are needed than in the case of large cell sizes. All the base stations must be connected by a wired communication network, such as the ones as presented in Sect. 2.3.

Additionally, more communication overhead results, because handoffs occur more often. A handoff names the process of a mobile node that changes its base station due to its movement: the context of a mobile node is switched to another base station in the case where the signal power falls below a certain threshold value.

During a handoff, a mobile node must establish a new communication channel to the new base station. Procedures like authentication must be performed, authorization must be verified, and so on. In addition, the availability of a communication channel and the required bandwidth to the new base station must be checked. Within a cell, for instance, time division multiplexing may be used, allowing multiple mobile nodes to communicate to the base station. If all time slices are already in use, no new connection can be established.

Besides connecting to the new base station, the mobile node must also disconnect from the old base station. To avoid a complete disconnection of the mobile node from all base stations during the handoff, there will be a short time during which the mobile node is connected to both base stations. This is feasible only if an overlap of the areas covered by the two base stations exists. Figure 2.37 shows such an overlap and a mobile node that crosses this area. If

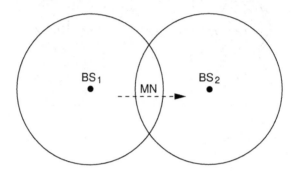

**Fig. 2.37.** Overlap of two base stations

the mobile node receives data from both base stations during a time interval, the handoff is called soft handoff. In contrast, a hard handoff switches from one base station to the other at a certain point in time.

Due to the overlap of adjacent base stations, as depicted in Fig. 2.37, adjacent base stations use different frequencies or different codes to avoid interference in the overlap area. In other words, frequency division multiplexing or code division multiplexing is applied.

To keep the amount of frequencies in the case of FDM as small as possible, base stations that do not overlap may use the same frequencies. Figure 2.36 shows how frequencies may be assigned to the base stations. A cluster includes all cells of equal frequency. Figure 2.36 consists of three clusters.

A further division of the area can be performed by partitioning a cell into slices. Sectorized antennas that transmit at different frequencies can be used to cover the different slices. If a cell suffers from much traffic, it can borrow frequencies from adjacent cells. This borrowing channel allocation (BCA) is much more flexible than the fixed channel allocation (FCA). If any frequency can be assigned to a base station, it is known as dynamic channel allocation (DCA).

### 2.4.3 Ad-hoc Networks

Cellular networks as described in the previous section need some infrastructure such as base stations. If no such infrastructure is available, for example, because it is too expensive or it takes too long to establish, mobile ad-hoc networks [39, 120, 127, 162, 174] may be used. This section gives an introduction, particularly to multi-hop ad-hoc networks.

Ad-hoc networks consist only of mobile nodes. A mobile node acts as the source node of a message, as its destination node, and as an intermediate node to forward the message to the destination or in its direction. Figure 2.38 depicts a possible scenario in an ad-hoc network. A mobile node would like to

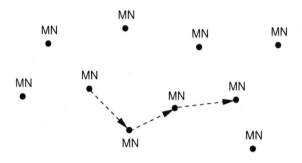

**Fig. 2.38.** Ad-hoc network

send a message to another mobile node. But because the source node is not able to directly reach the destination node, it first transmits the message to an intermediate mobile node which is in its range. This intermediate node sends the message to another intermediate node, which is able to directly reach the destination node and delivers the message.

One of the most important problems of ad-hoc networks is the routing, i.e., finding the right path to the destination node. The simplest way is the one in which each mobile node periodically sends a signal. An adjacent node receives this signal and updates its local table, which contains all neighboring nodes. This can be used by a sender, which sends the message to its neighbors according to its local table. Next, these neighbors send the message to their neighbors, and so on. Additional effort may be needed to avoid loops, to consider a fast changing topology, to deal with asymmetric links (a node may receive its neighbor but not vice versa), and so on. Because this book addresses more the network architecture and less the routing, the interested reader is referred to [158, 174] for routing details.

## 2.5 Network-on-Chip Architectures

In addition to wired and wireless network architectures, a third kind of network architecture can be distinguished that connects the components on a single chip. Currently, a bus topology usually provides the basis for such an architecture. But, in future, more complex architectures will be needed most likely. They are called network-on-chip (NoC). For instance, networks-on-chips [15, 46, 47, 187] are proposed to connect the components of a system-on-chip or the cores of a multicore processor. Such network architectures are also wired, but they must cope with particular constraints. In the following, the term network-on-chip also includes the bus topology. The term simply denotes all kinds of network architectures that are used for linking the components on a single chip.

### 2.5.1 Origin and Use

The ongoing improvement in very large scale integration (VLSI) technology leads to a further increasing number of devices per chip. Since this increased density can no longer be used to improve the performance of the components on the chips as in the past, the freed chip area can be used for other system components or for multiple processor cores in case of a former uniprocessor chip architecture [46].

### System-on-Chip

Due to the incorporation of other system components, an entire system may be built on a single chip, including digital and analog devices. Such a system is called system-on-chip (SoC). To shorten the development time of new chips, system components sometimes are prefabricated. The developer chooses from the available prefabricated components called intellectual property cores (IP cores), synonymous with intellectual property blocks (IP blocks). These IP cores may be soft IP cores given by software libraries or hard IP cores with positioned hardware elements and routed wires [135]. In addition to IP cores, entire processors, analog devices, and so on are also options for system components.

Figure 2.39 depicts a system-on-chip with its network-on-chip connecting some IP cores. SoCs may be complete computer systems, processors with many peripheral functions, or specialized systems for embedded applications. They have several advantages compared to multiple chip systems. Due to short distances between the components, SoCs operate much faster. The integration on a single chip provides lower assembly cost and higher reliability. The high integration of components also reduces the power consumption of each component (while the increasing number of components raises the overall power consumption).

Three different technologies are applied to realize a system-on-chip. These technologies include the full-custom realization, the standard cell realization, and the realization with field programmable gate arrays (FPGAs). The full-custom SoC sets up a chip with optimized circuits for a particular function that the developer is willing to realize. The standard cell SoC design is based

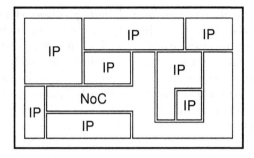

**Fig. 2.39.** System-on-chip

on a hardware description language (HDL) like VHDL or Verilog. These languages use standard libraries providing the logical and physical representations of standard components (IP cores). The SoC hardware realization can be performed by application-specific integrated circuits (ASICs). The third system-on-chip realization is by using FPGAs. FPGAs will be described in detail in Sect. 2.6.

### Multicore Processors

If the improvement in VLSI affects uniprocessor chips (with a single processor core), the chip area gained can be used to incorporate additional processor cores. Such a system is called multicore processor or multiprocessor system-on-chip (MPSoC). Currently, much less than ten cores is a reasonable number for a chip.

Figure 2.40 shows a multicore processor with its network-on-chip connecting the cores. Like multiprocessor systems such as parallel computers, multicore processors also outperform single core uniprocessor systems if a parallel execution of the applications is possible. A high speed-up in computing can be achieved. Despite some common features between them, multicore processors

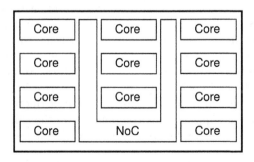

**Fig. 2.40.** Multicore processor

differ from multiprocessor systems with discrete processor chips in many ways. For instance, because chip area is limited, the cores of a multicore processor have usually no local memory (except a cache). All cores access a shared memory, which may be the bottleneck of the system as the number of cores grows. Some multicore processors even share a second level cache, causing the same problems as a shared main memory. On the other hand, the cache coherency circuits for first-level cache and (non-shared) second-level cache can operate at a higher frequency due to much shorter cache distances.

Concerning the level of parallelism, multicore processors are operating using thread level parallelism (TLP), which is also called chip-level multiprocessing (CMP) in this context.

Multicore processors are called symmetric if all cores are identical. If the cores differ, the multicore processor is asymmetric.

### 2.5.2 Particular Characteristics

For multicore processors, as well as for systems-on-chips, one of the main challenges is to find an appropriate network for connecting the (processor or IP) cores: the network-on-chip. In principle, the same networks as for off-chip connections can be applied. But NoCs reveal some particular characteristics that have to be considered [15, 171].

Comparing networks-on-chip with off-chip networks, one of the main differences is the reduced overall network area. Thus, a network that is set up on a chip results in much shorter distances between the network components. In consequence, signals on the wires reach their destination much faster than in off-chip networks. The NoC can exchange signals at a higher frequency. The clock cycle time of the network can be reduced significantly. On the other hand, a fast clocking in combination with global wires across the chip may lead to synchronization problems: a global synchronization may fail because the signal delay for traversing the chip takes more time than a clock period. Repeaters in long wires keep the delays linear to wire length instead of quadratic.

The signals on the shorter wires in NoCs require less voltage to be driven. Thus, less power is consumed by each network component, compared to an identical component of a previous chip generation, which is larger in size. But, on the other hand, the greater integration leads to more components on the same chip area. This counters the previously mentioned power saving properties, and the overall power needed increases. Additionally, it is also increased by the higher clocking on the chip. The growing power consumption implies that thermal problems may arise. Furthermore, increased power demands make power supply to the components on the chip difficult. Therefore, networks with basically low power demands are a good option for NoC architectures.

A way to save power in the network is by the reduction of network components in number and size. Fewer and smaller components consume less power.

Such networks additionally save chip area, which is expensive. The chip cost depends on the chip area by the power of three [83]. A reduced network chip area can be achieved by using simple components. For instance, NoC routers with reduced intelligence concerning network failures, routing alternatives, arbitration, and blocking behavior save logic circuits and, thus, chip area.

The high cost of chip area is also the reason for high memory cost. Memory requires a large chip area compared to other components. Therefore, fast static memory cells are particularly expensive. Smaller buffers should be chosen in NoC architectures. For instance, if a communicating pair of nodes always uses the same network path, no out of order sequences of data occur. A buffer to reorder data can be omitted. Furthermore, the switching technique of wormhole switching should be applied. Then, a buffer size to accommodate flits instead of packets is sufficient. If packet switching is not a precondition, circuit switching may be a better option than wormhole switching. No buffers are needed at all. But then, blocking of messages can be a problem.

Buffers should be more centralized. If buffers of relatively small size are distributed in the NoC, the overhead areas of the buffers accumulate uncontrollably. Internal buffering in crossbars must be avoided, for instance. An NoC architecture with a central buffer should be preferred to architectures with distributed buffers.

Because the network with its attached cores is located on a single chip, two advantages result. First, sorting out malfunctioning chips after chip production and testing lead to chips that rarely fail while operating. In consequence, any logical circuits to detect and handle failures of network components can be omitted in such reliable networks. Protocols need no error correction or retransmission features. This again saves chip area. Nevertheless, in the case of a poor chip design or bad environmental conditions, errors may occur due to the low chip voltages. Then, crosstalk, electromagnetic interference, and radiation-induced charge injection are not negligible in relation to chip voltage. Error correction logic can only be omitted if electrical noise can be kept within maintainable limits by using very conservative circuits [47].

In addition, the chip designer should be aware that network congestion may occur similarly to off-chip networks. Dropping data would then nevertheless require retransmission management, which is costly. A careful buffer implementation is usually a better way to address congestion. For the same reason, only deadlock-free network architectures should be considered for NoCs. For most topologies, deadlock-free routing algorithms are available.

The second advantage of single-chip systems is that the network with its attached cores needs no external pins to connect the network to the cores, neither on the network side nor at the cores. Network and cores are connected on-chip. Only pins for external devices must be incorporated, so less area for pins is required. Furthermore, the number of links per core is not as critical as for nodes in an off-chip network where each link to a neighboring node requires pins and powerful signal drivers. As a result, each core on the chip can be connected to many neighbors if allowed by the chip layout and the resulting wire crossings. In addition, much wider links than in off-chip networks can be introduced. For instance, 300 wires in parallel to simultaneously transfer 300-bit data are feasible [47, 105].

### 2.5.3 Topologies

Many topologies for networks-on-chips are feasible. In principle, all networks introduced in Sect. 2.3 are also appropriate for NoCs. But there are some constraints and specifics due to the characteristics presented in the previous section. They influence the NoC topology and other properties of the NoC architecture.

Two problems were pointed out: the power consumption problem and the area consumption problem. The power consumption problem leads to the conclusion that networks with basically low power demands should be preferred to NoC architectures. For instance, topologies that require very long wires across the chip must be avoided.

The area consumption problem favors networks with components of low number and small size. Networks that offer only simple functionality fulfill these requirements. As mentioned in Sect. 2.5.2, the single-chip property allows simple functionalities, for instance, as in case of the failure detection, which can simply be omitted. Furthermore, buffer sizes must be reduced and centralized as much as possible.

Besides these restricting features of NoCs concerning their architecture, there are also relaxing features, as derived in previous section. For instance, the number of links and wires per core (which is the communication node of the NoC) is not as critical as for off-chip networks. Thus, not only are network topologies with a low number of links like meshes or multistage interconnection networks a good option, a highly parallel data transfer is also feasible.

Currently, most system-on-chip and multicore processors use bus or crossbar topologies. Figure 2.41 depicts a current multicore processor consisting of two cores connected by a crossbar. Both cores of this dual core processor contain a CPU with its first level and second level cache. The crossbar connects them and provides a high speed connection to the (off-chip) main memory. If only two cores are involved, a crossbar or a bus is sufficient for the data exchange.

Busses are also appropriate for current SoC implementations. Their functionality and their low number of components only causes moderate network traffic that can be handled by busses. More complex networks are not yet needed.

But as the number of SoC components or the number of cores on a multicore processor grows [66], busses fail due to their bad blocking behavior. Furthermore, the power consumption of busses also increases because higher voltages are needed to drive the longer wires [20]. Crossbars must also be avoided due to the resulting size and the chip area consumption involved. Re-

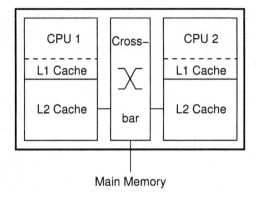

**Fig. 2.41.** Dual core processor with crossbar NoC

placing crossbars by multiplexers and demultiplexers can reduce this problem, but not completely solve it.

In the literature, many topologies are proposed for networks-on-chips. One of them is the multistage interconnection network topology. Guerrier and Greiner [73] established a bidirectional MIN structure (equivalent to a fat tree) using FPGAs. They called this on-chip network, with its particular router design and communication protocol, Scalable, Programmable, Integrated Network (SPIN). The network operates with a wormhole switching technique and with deterministic routing, although alternative paths exist in a bidirectional MIN. Its performance for different network buffer sizes was compared.

Alderighi et al. [3] used MINs with the Clos structure. Multiple parallel Clos networks connect the inputs and the outputs to achieve fewer blockings. Again, FPGAs serve as basis for realization.

Lahiri et al. [106] evaluated bus and ring topologies for NoCs. They investigated the performance of particular architectures of busses and rings dependent on spatial localities. Shared and hierarchical busses are used.

Shared busses also build the basis of Wingard's research [234]. His SoC communication infrastructure, called STBUS, can additionally be used to set up crossbars.

Wiklund and Liu [231] proposed a mesh based network-on-chip. They called it SoCBUS. It was developed especially for hard real-time embedded systems. Packets lock circuit parts while passing them.

Another mesh NoC was developed by Kumar et al. [105]. Their project describes a design methodology for generating the mesh architecture, and, in a second step, the application is mapped onto the mesh.

Lee et al. [112] presented a star network architecture. It is realized as a globally asynchronous system. They compared to it several other topologies.

Lüdtke et al. [124] investigated bidirectional MINs to be applied for reconfigurable NoCs. Particularly, profiting from the locality of bidirectional MINs was the goal of this research: a rewiring of the interconnections when the locality of network traffic changes heavily improves the network performance.

Sánchez et al. [173] described a reconfigurable direct network. In a two-dimensional torus topology, a node is able to exchange its position with a neighboring node.

Majer et al. [125] developed a simulator for evaluating a reconfigurable mesh architecture for NoCs, called DyNoC. The DyNoC is an incomplete mesh. Large chip areas are used for function blocks (Sect. 2.6), prohibiting mesh nodes from these areas.

More research has been performed in the area of network topologies for networks-on-chips. For instance, the reconfiguration of NoCs has been investigated by some researchers. Some of previously presented research papers deal with NoC topologies in combination with dynamic reconfiguration.

To support the design of a network-on-chip for a particular application, some tools have been developed. They help chose a feasible network architecture and offer some assistance in hardware development. To map the communication demands of the cores onto predefined topologies like meshes, MINs, and other topologies, Bertozzi et al. [17] developed a tool called NetChip (consisting of SUNMAP [143] and xpipes [16, 187]). This tool provides complete synthesis flows for NoC architectures. The authors investigated several topologies with their tool: mesh, torus, hypercube, and MINs.

Ching et al. [35] introduced a high-level NoC description language that eventually creates VHDL code. The related tool is also able to simulate the high-level description. Cycle-accurate performance results are obtained.

Table 2.3 gives an overview about proposed NoC topologies and related features. To summarize, previous publications propose the following topologies for NoCs: busses, meshes, tori, rings, stars, hypercubes, crossbars, and multistage interconnection networks. All of them were presented in Sect. 2.3.

## 2.6 Network Reconfiguration

Previous sections presented several network architectures. Usually, the optimal network architecture heavily depends on the applications running on the network nodes. Different kinds of applications produce different kinds of network traffic pattern and, thus, different kinds of network architectures deal with them best [106]. In consequence, a system with a changing network architecture dependent on the applications and the traffic may perform better than a system with a fixed architecture if the reconfiguration overhead does not counteract the performance gain. Such a reconfiguration of wired network architectures or network-on-chip architectures will be the topic of this section [11, 38, 224].

**Table 2.3.** NoC topologies

| Topology | Features |
|---|---|
| bus | hierarchical busses [106], STBUS [234], tool support by NetChip [17] |
| mesh | locked circuit parts by packets (SoCBUS) [231], mapping of applications [105], reconfigurable mesh [10, 125] |
| torus | neighboring nodes may be exchanged [173], tool support by NetChip [17] |
| ring | considering spatial localities [106], tool support by NetChip [17] |
| star | asynchronous realization [112] |
| hypercube | tool support by NetChip [17] |
| crossbar | composed using STBUS [234] |
| MIN | deterministic routing [73], FPGA realization [3, 73], reconfigurable bidirectional MINs [124], tool support by NetChip [17] |

### 2.6.1 Reconfiguration Types and Levels

Different types of network reconfiguration are distinguished: configuration, reconfiguration, and dynamic reconfiguration.

The term network configuration relates to a network that is composed of predesigned components. These predesigned components build the basis of the network architecture. The network architecture is set up by configuring the components and adapting them to fulfill the desired network features. Due to the predesigned components, network design becomes fast and easy. Only the component parameters have to be fixed. The network is configured during set-up, also called compile time. Therefore, the configuration is sometimes called compile-time reconfiguration (CTR). The term network reconfiguration describes the case where a network can be configured multiple times. That means that it can still be configured after network set-up, for instance, by changing network parameters: reconfiguration takes place.

Finally, dynamic reconfiguration of a network refers to a network reconfiguration that takes place while the network is operating. This means that parameters of the network are changed without interrupting the network in its task of continuously transferring messages. Because the reconfiguration is activated at run-time, it is also called run-time reconfiguration (RTR). Table 2.4 summarizes the meanings and differences of the three terms.

All kinds of network reconfiguration can be related to software and hardware. Software reconfiguration from the technical point of view can easily be applied by changing or updating the software of the network control units if

**Table 2.4.** Term overview

| Type | Characteristic |
|---|---|
| configuration | network parameter set-up at compile-time |
| reconfiguration | network parameter set-up after compile-time |
| dynamic reconfiguration | network parameter set-up at run-time |

the software is stored in erasable and rewritable memory, like RAM cells or flash memory. Software reconfiguration is a well-known issue and, thus, it is concentrated on hardware reconfiguration in the following.

Hardware reconfiguration can address different abstraction levels. At the gate level, simple switches or multiplexers can be reconfigured by changing the state of particular transistors. At the transfer level, crossbars, busses, and other network components are affected by the reconfiguration. Their components, like the multiplexers, are subject to change. And at the architecture level, the network topology, the switching technique, etc. are the objects of interest to be reconfigured. Of course, a reconfiguration at a higher level usually also causes a reconfiguration at the lower levels. In the remaining part of this section, the higher hardware levels of reconfiguration are considered. Particularly, the challenges of dynamic reconfiguration are presented.

### 2.6.2 Dynamic Reconfiguration

As previously mentioned, the performance of a particular network architecture heavily depends on the network traffic pattern. In consequence, if traffic patterns change, a dynamic reconfiguration of the network architecture to adapt it to the new traffic pattern may significantly improve the network performance if the reconfiguration overhead does not counter the performance gain (which will be discussed later in this section).

Figure 2.42 gives an example. Two applications are executed on the nodes of a multiprocessor system. The multiprocessor system consists of eight nodes, which are connected by a three-stage bidirectional multistage interconnection network with 2×2 switching elements.

Each application launches two processes. The processes of the first application are running on nodes P2 and P3, those of the second application on nodes P4 and P5 (Fig. 2.42(a)). The processes that belong to the same application are obviously communicating much more with each other than with processes belonging to different applications. The figure depicts the main communication paths.

Now, a third application is started that consists of four processes. The four available nodes, P0, P1, P6, and P7, are allocated by these processes. Figure 2.42(b) shows that their communication path is extremely long if P0 or P1 wants to exchange messages with P6 or P7. Messages must pass all stages from left to right and back again. A dynamic reconfiguration of the network

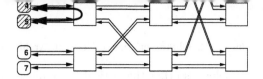

(a) Two running applications

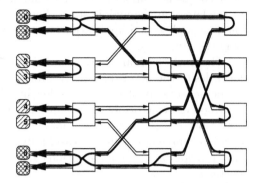

(b) New application started

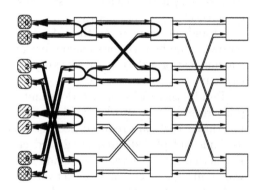

(c) Reconfigured network

**Fig. 2.42.** Dynamic network reconfiguration

topology, so that the critical communication path is shortened, improves the performance [124]. Changing the interstage connections in front of the first stage dynamically reconfigures the bidirectional MIN (Fig. 2.42(c)). The processes of the new application can communicate by messages that turn around, at the latest, at the middle stage. The lengths of the communication paths of the first and second applications are not changed. Their messages still pass only the first stage.

The previous example gave a scenario that profited from the topological localities of the network. The dynamic reconfiguration mapped the localities of the network topology onto the localities of the network traffic pattern.

Another scenario for a dynamic reconfiguration can be the changing of traffic patterns, resulting in a hot-spot node. The hot spot may appear due to a new application. In this case, a dynamic reconfiguration of the network that provides more communication capacity to the hot-spot node reduces this bottleneck.

Generally, dynamic reconfiguration can overcome temporary bottlenecks concerning the communication capacity. This can be achieved by temporarily dropping network characteristics that reduce performance. For instance, error checking in messages can be stopped. The time for error checking is saved and messages reach their destinations faster. In addition, the related hardware can be used for some performance-enhancing tasks. In this scenario, communication safety is sacrificed for the benefit of higher communication capacity.

Besides these self-optimizations and adaptations of network architectures, other reasons for network reconfiguration exist. For instance, design and manufacturing costs are reduced because predesigned network hardware can be used instead of dedicated integrated circuits, which need to be developed. Furthermore, new developments and standards can be incorporated into networks after their completion: networks with a long lifetime can be updated without exchanging hardware modules. Even changes during the design phase due to new standards or requirements can be handled without a complete network redesign.

Network hardware resources can efficiently be used by reconfiguring network components with different functionalities that are not required in parallel. This means that a network component can switch between two or more functionalities from time to time. Finally, reconfiguration can help increase fault tolerance and reduce failures. If parts of network components fail, important functionalities can be transferred to the remaining parts of the components in question.

Previous items point out the benefits of network reconfiguration. However, applying dynamic reconfiguration raises several questions and problems. For instance, packet switching may lead to a situation where packets are stored in the network buffers while the network is dynamically reconfigured. If this reconfiguration also changes the interconnections of the network (Fig. 2.42(b) and 2.42(c)), the stored packets must be handled with care: their path to the destination node may have changed or, worse, there may be no longer a path from the current position to the destination node due to the reconfiguration. Thus, a technique must be developed for dealing with such stored packets that ensures that they or some retransmitted copies reach the desired destination node. This technique must not reduce the network performance and counter the benefit of reconfiguration.

The dynamic reconfiguration will result in a new network architecture. For instance, the parameters of the network topology are changed in the above example. Thus, determining the new architecture is also related to the problem of packets stored prior to reconfiguration. On the one hand, the new architecture chosen should support the redirection of the stored packets. On the other hand, the reason for initiating the reconfiguration, e.g., changed network traffic patterns, must also be addressed. The challenge is to find an optimal new network architecture that is a good choice to act as the result of the dynamic reconfiguration.

Another important open consideration in this environment is the prediction of network traffic patterns of the near future. A dynamic reconfiguration of the network takes some time during which no messages are transferred. The time interval needed for reconfiguration is called reconfiguration phase. The length of the reconfiguration phase depends on the technology used. In the case of FPGAs, it may last from microseconds to a few seconds [135]. This means that a dynamic reconfiguration makes sense only if the operation phase is much longer than the preceding reconfiguration phase. The operation phase describes a time interval in which no reconfiguration occurs. In consequence, prior to a reconfiguration, it must be determined whether the reconfiguration is currently profitable. To do so, the network traffic of the near future must be predicted. Such prediction is usually very difficult.

### 2.6.3 Reconfigurable Hardware Architectures

For a dynamic reconfiguration of network architectures, some hardware is needed that provides this functionality. This section presents three hardware architectures supporting dynamic reconfiguration: FPGA, FPFA, and FPID.

### FPGA

Field-programmable gate arrays (FPGAs) are semiconductor chips that contain an array of function blocks, interconnection links, and input/output (I/O) blocks [38, 135]. Each of these components is reconfigurable. The structure of an example FPGA is depicted in Fig. 2.43. Many FPGA chips additionally provide separate memory blocks (also called block random access memory, BRAM). Sometimes, they also consist of embedded adders, multipliers, or even processor cores.

The interconnection links are organized into horizontal and vertical parallel wires (also called routing channels), with switch blocks at their intersections

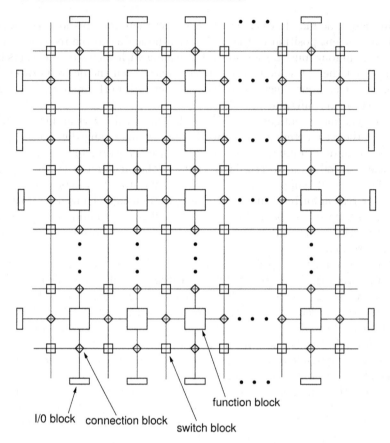

**Fig. 2.43.** FPGA structure

and connection blocks for attaching function blocks. Short wires can be used to set up local connections linking neighboring function blocks, while global connections result if wires are involved that pass the entire chip length or width. Many different wire lengths exist. Switch blocks and connection blocks link different wires together with switches to form a connection. Such connections can be changed by reconfiguring the switch blocks and connection blocks; the switch settings are then altered.

The I/O blocks are responsible for providing connections outside the FPGA chip. For this, they are linked with the external pins of the FPGA. Due to several different standards (e.g., of voltages) of the electrical signals that may occur outside the FPGAs, an interface is required that can be configured and adapted to a particular standard. I/O blocks fulfill this task. Some FPGAs additionally consist of some gigabit transceiver blocks for high-speed data exchange.

The function blocks (also called logic blocks, configurable logic blocks, or CLBs, or logic array blocks, or LABs) can be reconfigured so that they represent some basic logic gate functionality like AND, OR, or NOT. FPGAs with these functional blocks are called fine-grained FPGAs. FPGAs that are denoted as coarse-grained FPGAs can handle more complex functionalities within a functional block, such as multiplexers, encoders, or mathematical functions needed for algorithms.

Function blocks usually consist of several logic cells (also called logic elements). An example of a logic cell is given in Fig. 2.44. It is called an LUT-

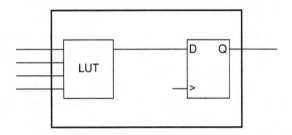

**Fig. 2.44.** Logic cell

based logic cell because the main component of it is a look-up table (LUT). In the LUT, dedicated return values are stored for each combination of the values of the $n$ inputs: a truth table is represented. The return value is delivered to the D-type flip-flop at the cell output where it is stored. Changing the return values of the LUT reconfigures the logic cell. In some FPGA architectures [135], LUTs are also able to store a $2^n$-bit value or to act as a $2^n$-bit shift register that provides additional functionality.

An alternative to LUT-based logic cells are MUX-based ones. In this case, multiplexers (MUXs) are the main components. Their inputs, including their control inputs, can be reconfigured such that a 0, a 1, or the binary output of another cell is assigned to them. Multiple multiplexers can be connected to realize a truth table, as in a LUT-based approach.

Nowadays, most FPGAs are LUT-based due to their higher speed. Their look-up tables are usually realized using static random access memory (SRAM) technology. SRAM allows infinite and fast reconfigurations. In contrast, technologies based on electrically erasable programmable read-only memory (EEPROM) and on flash memory are slower, and technologies based on anti-fuse links are configurable only once.

The task of reconfiguring an FPGA is called programming. Before programming, an FPGA design must be determined that shows the desired chip behavior. In the first step, this behavior is usually specified via a hardware description language (HDL). To accelerate the development, the HDL often provides libraries with predefined components. As already mentioned, they are called soft intellectual property blocks (soft IP blocks).

If functional simulation shows the correctness of the HDL specification, synthesis maps it to a netlist that is used to generate a gate-level description. A further simulation at this level checks the correctness again. Mapping the gate-level description to the FPGA elements gives the signal delay times. They are incorporated into the simulation for a last check. FPGA manufacturers provide tools supporting many of these development steps. These tools also perform the last step in design: they eventually generate the configuration file (also called bit file) that is uploaded into the FPGA to program it. While uploading, the configuration file is called configuration bitstream.

The configuration file for SRAM-based FPGAs consists of configuration data and configuration commands. The configuration data describes the new states of the reconfigurable FPGA devices. The configuration commands define how to handle the configuration data. Configuration cells on the FPGA chip realize the configuration of the FPGA blocks. These cells are addressed like a long single chain of cells. Using configuration port programming and serial load mode, the configuration file is transferred bit-serial from an external memory to the FPGA. There, the bits are shifted through the configuration cell chain. When the bits reach their corresponding configuration cells, the FPGA blocks can be configured.

A parallel load mode is also available. If applied, eight bits (i.e., a byte) are transferred in parallel from memory to the FPGA. There, the bits are loaded either in a serial manner with high speed into the configuration cell chain or in parallel into a system of eight chains.

Multicontext FPGAs allow storing configuration bits of multiple configurations in the configuration cells [38, 198]. In the case of switching between only a few configurations, they can be, and configuration bits need not to be transferred at each reconfiguration.

A reconfiguration while operating (that is a dynamic reconfiguration) is supported by many FPGAs. If FPGA blocks are reconfigured with the same data as in the preceding reconfiguration, the blocks do not change, and operate without interruption. FPGAs that are dedicated for dynamic reconfiguration keep the contents of their registers during the reconfiguration phase.

Another feature of some FPGAs is their capability of partial reconfiguration. Single columns of FPGA blocks can be reconfigured separately. The advantage is twofold. First, bitstreams are of smaller size and, thus, FPGA programming is accelerated. Second, the non-reconfigured parts are not involved, and can continue operating unpersuaded.

Reconfiguring parts of the FPGA while other parts keep operating helps overcome one of the main drawbacks of reconfiguration. Reconfiguration takes some time, as described in the previous section. During the reconfiguration phase, the FPGA will be in an undefined state (at least for those regions where a change takes place). In consequence, this time is lost, for example, for transferring messages if the FPGA acts as a network. A reconfiguration is advantageous only if the performance benefit of the reconfiguration overbalances the drawback of transfer interruption. In the case of partial reconfiguration, transfer interruption can be avoided if a way is found to reconfigure only parts of the network that are currently inactive. But one should be aware that this may not always be possible.

A technique called pipeline morphing [224] is sometimes proposed for dynamically reconfigurable FPGAs that act as processors. It addresses computations that can be divided into multiple steps according to pipeline stages. If there are too many stages to be mapped to an FPGA, the stages are successively realized by dynamic reconfiguration of the FPGA. Transferring this technique to FPGAs that act as networks does not seem feasible. Due to the reconfiguration phase, message delay times, a very important network performance measure, would be increased significantly.

However, FPGAs are a weak option to realize dynamically reconfigurable network architectures [38, 173]. Due to the fixed horizontal and vertical wiring of FPGAs, the network topology can only be changed in a very limited fashion. New FPGA architectures are needed [11, 124, 125]. On the other hand, network parameters like buffers can be reconfigured very well with current FPGAs.

**FPFA**

Field-programmable function arrays (FPFAs) represent a variation of FPGAs. Sometimes, FPFAs are also called field-programmable nodal arrays (FPNAs) [135]. Compared to FPGAs, they are not organized bit by bit but word by word. This data organization better supports complex data computations. Multiple-bit words are processed by complex function blocks [10]. Thus, FPFAs are usually coarse grained. This coarse-grained structure leads to a greater power efficiency, compared to fine-grained FPGAs.

The function blocks of FPFAs are sometimes called reconfigurable data path units (rDPUs). Often, they are simple reconfigurable processing elements. An example of an FPFA structure is, for instance, the reconfigurable data path architecture (rDPA) [78].

There also exist chips with a mixture of bit by bit data organization and word by word data organization. They were referred to as field-programmable mixed arrays (FPMAs). FPMAs are called multi-grained.

**FPID**

Field-programmable interconnect devices (FPIDs) are based on the same technology as FPGAs. FPIDs are also called field-programmable interconnect chips (FPICs). Their task is to act as chips that connect other chips or components by offering short delays [93, 117, 227]. Figure 2.45 shows an example where some FPIDs are distributed between other chips on a board. The FPIDs can be dynamically reconfigured to realize any desired connec-

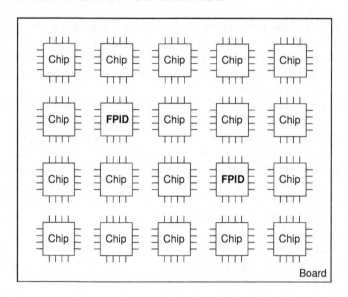

**Fig. 2.45.** Board with two FPIDs

tions. This means that any of their pins can be internally connected to any other pin.

In future, combining FPIDs and FPGAs on a multichip module (MCM) or on a system-on-chip device will further increase the flexibility in designing dynamically reconfigurable network architectures.

# 3

## Performance Evaluation

High performance is the most important goal in designing parallel or distributed systems. Therefore, performance evaluation is needed to compare various architectures for their performance.

Three kinds of performance evaluation methods can be distinguished: measurement at an existing system, numerical simulation, and mathematical methods. Performance measurement at an existing system (physical measurement) suffers from the drawback that the system must first be established in hardware before any performance evaluation is possible. If it turns out that this system architecture shows low to bad performance, new hardware must be realized, usually generating huge cost.

Another disadvantage of physical measurement is due to self-interference. Any measurement device somehow changes the system set-up, and therefore, the system behavior changes compared to a set-up without such a device. In consequence, system performance may also be influenced. Measurement is, however, important in order to verify that a realized system performs according to specified requirements, and that the design and implementation process has been successful.

The focus in this book is on simulation and mathematical methods used during the design phase of the system. These methods evaluate the performance of the system in question by first setting up a stochastic model for it. Then, the performance measures of the model are determined. Stochastic models are needed in order to include non-deterministic events like network traffic generation in space and time.

The mathematical methods applied include Markov chains and Petri nets. Other methods such as queuing theory with single station queuing systems or queuing networks are not used in this book, and thus will not be presented here. Detailed information about queuing theory can be found in [19, 61, 72, 95, 96, 113, 137, 199].

## 3.1 Numerical Simulation

Numerical simulation is based on the concept of designing a model of the system in question. This model describes the system behavior either by its discrete events (using the discrete event system specification, or DEVS, formalism), by its discrete time behavior (using the discrete time system specification, or DTSS, formalism), or by its continuous time and state change (using the differential equation system specification, or DESS, formalism) also referred to as continuous simulation.

Continuous simulation [248] can be applied if the state variables of the system under consideration change continuously in time. Then, the model description is based on differential equations. Often, ordinary models result in a system of differential equations that is too large to be solved symbolically. Less complex systems sometimes can be solved numerically to determine the steady state and the desired performance results.

The steady state describes the system state in which no further change occurs. A steady state $\bar{\theta}$ of a measure $\theta$ exists if the following equation holds in time $t$:

$$\theta(t) = \bar{\theta} \text{ for } t \to \infty. \tag{3.1}$$

Discrete time simulation [248] can be applied if the state variables of the system in question change only at discrete points in time. The simulator determines the system state each time step. For instance, it counts how frequent a particular state is reached. Often, applications for discrete time simulation are digital systems that are clocked. The discrete time steps are given by the frequency of the clock.

In discrete event simulation [9, 57, 88, 107, 248], the system is modeled by its events. Starting in a particular state, the model determines the next state transition that occurs. The new state is calculated. The measures of interest are observed by considering how long this state lasts. This means that the time until the next event occurs (state transition) is determined. This scheme is continued until the steady state is reached or any other termination criterion is fulfilled.

Discrete event simulation that includes stochastic events is sometimes called Monte Carlo simulation in literature. This name historically comes from random number driven methods of numerical mathematics, and does not exactly apply to stochastic discrete event simulation as used in this book. As already mentioned, times between different states are observed for a dynamic process [107], while for Monte Carlo simulations, events can be described as a simple function of random numbers [172]. Determining events of a parallel or distributed system may require very complex functions or algorithms, and cannot be called a Monte Carlo simulation.

### 3.1.1 Statistics

Simulation is usually applied to determine an unknown performance measure $\theta$ in a random variable $Y$: $\theta = E(Y)$. Several realizations $(Y_1, Y_2, \ldots)$ of this random variable are generated. A single realization $Y_i$ is also called observation, and represents a sequence of state transitions. A subset of reachable states is passed. In statistics, an observation is also called sample. Because a sample or observation represents a time series by a tuple of single values and their occurrence in time, there are statistical methods available to characterize the observations.

### Mean and Accuracy Estimation

If all $Y_i$ result from the steady state, the expected value $\mu = E(Y)$ (also called expectation) can be estimated by the arithmetic mean

$$\hat{Y}(n) = \bar{y} = \frac{1}{n} \sum_{i=1}^{n} Y_i. \tag{3.2}$$

It is also important to get an idea of how close this estimated value is to the real value. Such a clue gives the variance, defined as the mean quadratic deviation of the random variable from the expected value:

$$\mathrm{Var}(Y) = E((Y - E(Y))^2) = E(Y^2) - (E(Y))^2. \tag{3.3}$$

For a sample of size $n$, an estimation of its variance is

$$\hat{\mathrm{Var}}(\bar{y}) = \frac{1}{n(n-1)} \sum_{i=1}^{n} (Y_i - \bar{y})^2. \tag{3.4}$$

The variance is also called second central moment. In general, the $k$-th central moment is given by $E((Y - E(Y))^k)$.

Equations (3.2) to (3.4) are only valid if all values are normally distributed and independent of each other. Particularly, the latter issue reveals a problematic condition often not fulfilled when performing multiple realizations of $Y$. Values within a single realization are usually highly dependent on each other. For instance, if a network is congested, messages sent through the network will show high delay times and influence the delay times of the following messages because they also contribute to network congestion.

The dependence between two values can be described by the autocorrelation

$$\rho(k) = \rho(-k) = \mathrm{Corr}(Y_i, Y_{i+k}) = \mathrm{Corr}(Y_i, Y_{i-k}), \tag{3.5}$$

where $k$ $(0 \leq k \leq n-1)$ is called lag. The autocorrelation can be determined by the covariance

$$\gamma(k) = \gamma(-k) = \mathrm{Cov}(Y_i, Y_{i+k}) = \mathrm{Cov}(Y_i, Y_{i-k}) \tag{3.6}$$

using the relation $\rho(k) = \gamma(k)/\gamma(0)$. The autocorrelation results from normalizing the covariance by the variance. Taking the autocorrelation into account, the variance is given by

$$\mathrm{Var}(\bar{y}) = \frac{1}{n}\tilde{R}\gamma(0), \tag{3.7}$$

with

$$\tilde{R} = 1 + 2\sum_{k=1}^{n-1}\left(1 - \frac{k}{n}\right)\rho(k) \tag{3.8}$$

as a function of the autocorrelation of the observed process, and $n$, the number of available results. In the case of completely independent values, $\rho(k) = 0$ holds for all $k \geq 1$, and therefore, $\tilde{R} = 1$. In the case of dependent values, and if an infinite number is available, $\tilde{R}$ approximates

$$R = \lim_{n\to\infty}\left(1 + 2\sum_{k=1}^{n}\left(1 - \frac{k}{n}\right)\rho(k)\right) = 1 + 2\sum_{k=1}^{\infty}\rho(k) \tag{3.9}$$

and denotes the number of correlated values that carry the same information as a single value of uncorrelated values.

If $R$ exists and is finite, the real variance, given as the sum of all covariances, results in

$$\sigma^2 = \lim_{n\to\infty} n\,\mathrm{Var}(\bar{y}) = \sum_{k=-\infty}^{\infty}\gamma(k) = \gamma(0)\,R. \tag{3.10}$$

An estimation of the variance in the case of a finite number of values then leads to

$$\underline{s^2} \approx \frac{1}{n}\left(1 + 2\sum_{k=1}^{n-1}\left(1 - \frac{k}{n}\right)\frac{c_k}{c_0}\right), \tag{3.11}$$

with

$$c_k = \frac{1}{n-k}\sum_{i=1}^{n-k}(Y_i - \bar{y})(Y_{i+k} - \bar{y}), \tag{3.12}$$

known as empirical covariance.

If correlated values exist and the standard estimation of Eq. (3.4) is used instead of that of Eq. (3.11), the calculated variance may differ significantly from the real variance.

## Spectral Analysis

One of the methods to deal with correlated observations is called spectral analysis. Two versions exist. The first is given by Fishman [56]. It allows an estimation of the variance if the process shows weak stationary behavior, defined as

- $E(Y_i) = E(Y)$ for $i \in \mathbb{N}\backslash\{0\}$ and $-\infty < E(Y) < \infty$
- $\text{Var}(Y_i) = \text{Var}(Y)$ for $i \in \mathbb{N}\backslash\{0\}$ and $\text{Var}(Y) < \infty$
- $\text{Cov}(Y_i, Y_{i+k})$ is independent of $i$ for $i \in \mathbb{N}\backslash\{0\}$

Then, the covariance $\gamma(k)$ and the spectral density $p(f)$ can be expressed as the Fourier transform pair

$$p(f) = \sum_{k=-\infty}^{\infty} \gamma(k)\cos(2\pi fk) \quad \text{and} \tag{3.13}$$

$$\gamma(k) = \int_{-\frac{1}{2}}^{\frac{1}{2}} p(f)\cos(2\pi fk)\, df. \tag{3.14}$$

Case $f = 0$ yields $p(0)$ as the sum of all covariances, i.e., the variance $\sigma^2$.

Equation (3.13) converts the covariances to the frequency domain, and variance estimation can be performed by estimating the variance of the spectral density at $f = 0$. Unfortunately, the precision of such estimation is usually very poor due to huge differences between the values that the function produces [107].

A second version of spectral analysis was described by Heidelberger and Welch [81]. This version is more appropriate than the one just presented. It is based on the periodogram

$$\Pi\left(\frac{j}{n}\right) = \frac{|A_y(j)|^2}{n} \tag{3.15}$$

of a series of observations $Y_1, Y_2, \ldots, Y_n$ and the discrete points $j$. $A_y(j)$ represents the fast Fourier transform of the series of observations $Y_k$ as

$$A_y(j) = \sum_{k=1}^{n} Y_k e^{\frac{-2\pi i(k-1)j}{n}}, \tag{3.16}$$

with $i = \sqrt{-1}$ giving the imaginary unit. An accurate estimation of the spectral density

$$p_Y\left(\frac{j}{n}\right) \approx E\left(\Pi\left(\frac{j}{n}\right)\right) \tag{3.17}$$

results and determines the variance of the time series for $\frac{i}{n} \to 0$. Smoothing this function facilitates variance estimation: the periodogram is logarithmized and two succeeding values are merged. The smoothened periodogram

$$L(f_j) = \ln\frac{\left(\Pi\left(\frac{2j-1}{n}\right) + \Pi\left(\frac{2j}{n}\right)\right)}{2} \tag{3.18}$$

of the frequencies $f_j = (4j-1)/n$ is approximated by a polynomial for further simplification.

This method estimates the variance for a sequence of observations. Accuracy can be determined via this variance (see Sect. 3.1.1). When new sequences are available, the newly calculated variances help sequentially improve accuracy until termination criteria are met.

## Batch Means

Besides using spectral analysis, the method of batch means is also often applied for variance estimation of correlated observations. This method divides the series of observations $Y_1, Y_2, \ldots, Y_n$ into $b = \lfloor n/m \rfloor$ sequences of non-overlapping groups of size $m$. Then, the first group consists of $Y_1, Y_2, \ldots, Y_m$, the second group of $Y_{m+1}, Y_{m+2}, \ldots, Y_{2m}$, and so on. Further statistical analysis is based on the means $\bar{y}_1(m), \bar{y}_2(m), \ldots, \bar{y}_b(m)$ of each group (Fig. 3.1, top). The method results from the idea that the more the time spent between

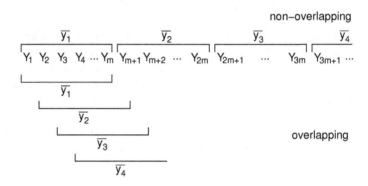

**Fig. 3.1.** Batch means

two observations, the less their correlation. Therefore, the group size $m$ should be as large as possible. On the other hand, if $m$ gets too large, many observations are needed to estimate the variance, and therefore, simulation takes a long time.

One of the methods to sequentially determine $m$ tests whether a given $m$ leads to only slightly correlated means and increases $m$ in the case of strong correlation [153]. Slight correlation is tested by the estimation of the empirical covariance $c_k$ (Eq. (3.12)) of the lag $k$. A very small ratio $c_k/c_0$ (e.g., less than 0.05) indicates slight correlation.

A modification of the previously described method of batch means uses overlapping groups [138]. Each observation starts a new group (Fig. 3.1, bottom). This means that each observation belongs to $m$ groups (if the group size is $m$). Then $n - m + 1$ groups of size $m$ result instead of the $\lfloor n/m \rfloor$ in the case of non-overlapping groups. The group size may be determined identically to the algorithm given in the case of non-overlapping groups. A drawback is a less detailed prediction about accuracy (e.g., a worse confidence level, as defined in Sect. 3.1.1) due to overlapping [138].

## Confidence

Previously determined variance now helps giving a confidence [169, 178] to the results. The confidence for a sample size of $n$ is expressed by the confidence level

$$P\left(\bar{y} - t_{n-1,1-\alpha/2} \cdot \sqrt{s^2/n} \le E(Y) \le \bar{y} + t_{n-1,1-\alpha/2} \cdot \sqrt{s^2/n}\right)$$
$$= 1 - \alpha \qquad (3.19)$$

with $\alpha$ giving the expected error and $t_{n-1,1-\alpha/2}$ giving the quantile (Sect. 3.1.1) of the Student's $t$-distribution with $n-1$ degrees of freedom and the expected error. Equation (3.19) holds only if the values determined by simulation are normally distributed. The interval within the brackets on the left-hand side of the equation is called confidence interval. For instance, if $k$ experiments containing $n$ realizations exist, $k$ different confidence intervals result. Then, $k \cdot (1 - \alpha)$ of them will include the expected value $E(Y)$.

Confidence is also often specified by the estimated precision (also called relative statistical error),

$$\varepsilon = \frac{\Delta y}{\bar{y}}, \qquad (3.20)$$

with

$$\Delta y = t_{n-1,1-\alpha/2} \cdot \sqrt{s^2/n} \qquad (3.21)$$

giving the absolute statistical error. This means that the estimated precision is closely related to the confidence level.

## Initial Transient Phase

During a single simulation run, two phases can be distinguished. In the first phase, called initial transient phase, the system model transiently oscillates until a steady state (if it exists) is reached. The steady state represents the second phase.

Some investigations aim to determine measures $E(Y|t = t_0)$ at a particular time $t_0$ (called terminating simulation) while others are interested only in the steady state $E(Y|t \to \infty)$ (called steady-state simulation). In the latter case, values of the initial transient phase distort the results, particularly the confidence level. Therefore, determining the initial transient phase and starting the observation of results in steady state improves the results and is the main task in steady-state simulation.

In [153], the main methods to estimate the length of the initial transient phase are presented. Most of them base on heuristics. One of the commonly applicable methods is the Schruben test [175]. The Schruben test does not directly determine the length of the initial transient phase. It tests for a part of the simulation run whether there is any transient influence or this part is in a steady state. It profits from the high sensitivity of partial sums in time series. Partial sums $S_n(k) = \bar{y}(n) - \bar{y}(k)$ are established by considering the mean of the first $n$ and $k$ values, respectively. The time series

$$T_n(t) = \frac{\lfloor nt \rfloor S_n(\lfloor nt \rfloor)}{\sqrt{n \, \hat{\mathrm{Var}}(Y(n))}} \tag{3.22}$$

can then be calculated, where $0 < t \leq 1$ and $T_n(0) = 0$. It is assumed that such a series converges to a Brownian bridge process. This process represents a mathematical model of Brownian motion (a random walk with random step sizes) in the interval $[0, 1]$. Testing this process for its steady state is more convenient than for most other processes [175].

Before this method of the Brownian bridge process is started, a heuristic is applied to accelerate the detection of the steady state: if a series $Y_1, Y_2, \ldots, Y_{n_0}$ crosses its mean $\bar{y}(n_0)$ for the $j$-th time, it is assumed to be close to its steady state [153]. Often, a value of $j = 25$ is proposed. Figure 3.2 illustrates the Schruben test. Simulation tools like *TimeNET* [65, 88] and *Akaroa* [155] detect the steady state using the Schruben test.

### Skewness and Quantiles

Estimating the mean of the performance measure in question (represented by a random variable) may lead to incomplete or wrong conclusions. The distribution of this random variable is also very important. It provides useful information about the deviation of the values. In Sect. 3.1.1, the variance was defined. It gives information about how close the realizations $Y_i$ are to the mean on average.

The skewness is another parameter giving information about the shape of the distribution. The skewness $\nu$ is defined as

$$\nu = \frac{E((Y - E(Y))^3)}{(\sigma^2)^{3/2}}. \tag{3.23}$$

It describes the symmetry of the probability density function or the probability mass function of the distribution. In case of $\nu = 0$, the distribution is symmetric. For instance, the normal distribution is a symmetric distribution. For $\nu > 0$, the distribution is skewed to the right, and for $\nu < 0$, it is skewed to the left.

Quantiles are also helpful for estimating the shape of the distribution. For a continuous distribution function $A(t)$ of a continuous random variable $t$ with $0 < A(t_1) \leq A(t_2) < 1$ for $t_1 < t_2$, the $q$-quantile $t_q$ is defined as

$$t_q = A^{-1}(q), \tag{3.24}$$

with $0 < q < 1$ and $A^{-1}$ representing the inverse function of $A$. This means that $t_q$ fulfills the equation $A(t_q) = q$.

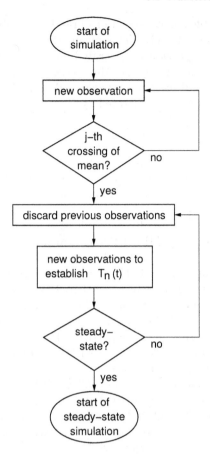

**Fig. 3.2.** Schruben test

The median, the quartiles, and the octiles are particular quantiles. The median denotes the 0.5-quantile. In the case of a symmetric distribution, the median is equal to the mean. The quartiles are the 0.25-quantile and the 0.75-quantile. The octiles are given by the 0.125-quantile and 0.875-quantile.

Techniques and methods for quantile estimation in steady-state simulation can be found in [80, 111].

### Random Number Generator

A random number generator is needed if stochastic events are modeled. Such algorithmic generators provide uniformly distributed pseudo-random numbers to include randomness in simulation [82, 154]. Many algorithms are proposed in the literature [97].

The unsolved problem in construction of random number generators is that they generate numbers in cycles. This means that after generating a certain quantity of numbers a previously generated number is generated again. The same sequence of numbers starts again due to the generation algorithms, which are usually recursive and based on integer modulo $M$ arithmetic. A series of random integer numbers $I_i$ is generated by

$$I_i = (a \cdot I_{i-1} + c) \bmod m, \qquad (3.25)$$

with the multiplier $a$, the increment $c$, and the modulus $m$ as non-negative parameters. The seed of the series is given by $I_0$. In the case of $c > 0$, the generators are called mixed linear congruential pseudo-random number generators.

Current generators are usually multiplicative linear congruential pseudo-random number generators (MLC-PRNG), which means that $c$ equals 0. For instance, the ANSI C rand() function belongs to this class.

Nowadays, computers with 32-bit architecture lead to a modulus of $M = 2^{31} - 1$. With clock frequencies of more than 1 GHz, all numbers of a cycle can be generated within a few minutes. Thus, simulation runs lasting longer than only a few minutes will deal with cyclically generated random numbers. Exhaustive simulation run times are particularly common in rare event simulation, for instance, if packet loss in the ATM environment is measured, or if self-similarity prevents fast confidence.

Some recently developed generators called multiple recursive LC-PRNGs and combined multiple recursive LC-PRNGs [109] overcome this problem and offer cycles between $2^{185}$ and $2^{377}$. Furthermore, generalized feedback shift register PRNGs also lead to reasonable cycle size. The Mersenne Twister [134] as an example provides cycles of $2^{19937} - 1$.

### 3.1.2 Acceleration

One of the main drawbacks of simulation models is their very high simulation run time. If models include stochastic events, run times may reach hours, days, or even weeks until the desired confidence level is reached. Real-world scenarios are usually influenced by stochastic events generated by the environment.

Therefore, many methods have been developed to accelerate simulation runs. In the following, the most important of these methods are presented. Two types can be distinguished: methods to reduce the variance and methods to enforce more relevant events.

### Variance Reduction

Accelerating the simulation by variance reduction means increasing the accuracy of the results and, thus, the confidence. Due to Eq. (3.19), the confidence level is increased by reducing the variance $\underline{s^2}$ or by increasing the number of observations $n$. In other words, if the variance is reduced, the number of observations can also be reduced while keeping the confidence the same.

Many methods for variance reduction can be found in the literature (for instance, [9, 23, 34, 57, 107, 147, 233]). Some of them will be briefly discussed in the following. They apply to discrete simulation.

If stratification is chosen to reduce the variance, the set of observations is divided into disjunct strata. The number of observations for each stratum determines the stratum size. A temporary variable $X$ is introduced representing a stochastic parameter of the simulation model such that each $X^{(i)}$ relates to a $Y^{(i)}$. $Y^{(i)}$ may be a single observation or a mean. It is assumed that the relation between simulation input and $X$ is known. Then, all probabilities $P(X = i)$ are also known. An estimation $h(Y)$ may now be decomposed to

$$h(Y) = \sum_{i=1}^{n} h(Y|X = i)P(X = i) \tag{3.26}$$

$$= E(h(Y|X)). \tag{3.27}$$

$Y|X = i$ gives $Y$ if $X = i$ holds. By determining the variance $\mathrm{Var}(Y)$, the equation gives the average variance of $Y$ for all given $X$. Additionally, the variability of $X$ must also be considered. The variance is given by [34]

$$\mathrm{Var}(Y) = E(\mathrm{Var}(Y|X)) + \mathrm{Var}(E(Y|X)). \tag{3.28}$$

Because stratification requires that all probabilities $P(X = i)$ are known, it reduces the variance $\mathrm{Var}(Y)$ by these probabilities (Eq. (3.26)).

An extension of stratification is called importance sampling [69], where stratified sampling is changed such that weights of the probabilities of $X$ are modified. A modification is chosen in a way such that most observations belong to high probabilities $P(X = i)$ and are determined for high conditional variances. But this also modifies the probability distribution of the model input and, thus, modifies the model.

Instead of assuming that the probabilities $P(X = i)$ are known, as in stratification, the conditional Monte Carlo method assumes that the expected values $E(Y|X = i)$ are known. Variance reduction is achieved by considering Eq. (3.28). But [23] shows that this assumption holds only for particular scenarios.

Many further methods to reduce the variance are known, such as control variables, antithetic variates, and Latin hypercube sampling. They introduce dependences or modify the observations or the sequence of observations. An overview is given in [9, 23, 57, 147].

**Enforcement**

In addition to variance reduction, enforcement also accelerates simulation. This technique either changes the model or influences the simulation sequence.

Enforcement aims to achieve more observations without extending simulation run time. Two main methods can be distinguished: parallelization of the simulation run and boosting the occurrence of the measured event.

A main issue in parallelization deals with dividing a single simulation model into several submodels to be executed on several nodes of a parallel or distributed computer system. Usually, the submodels will somehow interact because they depend on each other [60]. If this dependence is too strong, the interaction results in heavy communication consuming much computing power. In consequence, parallelization does not accelerate the simulation time very much, or even retards it [26]. A high performance communication network connecting the nodes is essential.

Another problem arises by the partitioning of the model into submodels. To automatically divide the model is usually not possible due to complex and varying models. On the other hand, manually dividing them consumes much time, and is reasonable only in the case of very long simulation run times or if the model is used for multiple investigations. Furthermore, all submodels should require similar computation power to ensure a balanced distribution of the entire model among the nodes.

Instead of parallelization of the model via submodels at the nodes, parallelization of the entire model also offers a way to accelerate the simulation. Multiple replications of the model are started in parallel on the nodes of a parallel or distributed computer system. This method, called multiple replications in parallel (MRIP) [88, 155], is based on stochastic behavior: many uncorrelated observations are needed to establish statistics about accuracy and confidence (see Sect. 3.1.1). Independent simulations in parallel supply such observations if all random number generators start with different roots. Nevertheless, convergence to wrong values can be a problem particularly if a large number of nodes perform the simulation [70]. Spectral analysis as accuracy estimation overcomes this problem [155].

Due to completely independent replications, no synchronization is needed. Communication occurs once for distributing the simulation model among the nodes. Further communication arises only from result statistics to determine current confidence levels and to observe termination criteria. Statistics can be determined by establishing points in time for each replication $r_i$, at which the amount of local observations $n_{r_i}$, local means $E(Y_{r_i})$, and local variances $\text{Var}(Y_{r_i})$ are calculated. Then,

$$E(Y) = \frac{\sum_i n_{r_i} E(Y_{r_i})}{\sum_i n_{r_i}} \tag{3.29}$$

gives the global mean and

$$\text{Var}(Y) = \frac{\sum_i n_{r_i}^2 \text{Var}(Y_{r_i})}{\left(\sum_i n_{r_i}\right)^2} \tag{3.30}$$

gives the global variance.

In addition to parallelization, boosting the occurrence of the measured events represents a method to accelerate simulation. It can particularly be helpful to measure rare events. As an example, a method called repetitive simulation trials after reaching thresholds (RESTART) [226] will be discussed here. This method profits from a more frequent occurrence of the rare event $A$ if simulation starts from a particular model state given by the event $C$ that enlarges the probability of the rare event $(A \subset C)$. This means that

$$P(A) = P(C) \cdot P(A|C), \tag{3.31}$$

with $P(A) \ll P(C) \ll 1$. RESTART improves the estimation of $P(A|C)$ by multiple repetitions of the simulation run starting at event $C$. Thus, more observations of the rare event result, and an accurate probability of it can be determined faster. Introducing a sequence of such intermediate events $C_i$, with $C_1 \subset C_2 \subset \ldots \subset C_m \subset A$, further accelerates the simulation. The RESTART algorithm is given as follows:

1. If the intermediate event $C_i$ occurs, the system state is saved.
2. The simulation run is started/continued from the saved system state.
   - If an event $D_i$ occurs that reveals a worse reachability of the rare event $A$ than $C_i$, the simulation run is stopped. The algorithm continues with step 2.
   - If the intermediate event $C_{i+1}$ occurs
     - and $C_{i+1} \neq A$, the new system state is saved and the algorithm continues with step 2 at the newly saved state.
     - and $C_{i+1} = A$, the rare event is reached. Statistics are modified. The algorithm continues with step 2.
   - If a fixed number $R_i$ of simulation runs starting at a system state related to $C_i$ is performed, the previously saved state (related to $C_{i-1}$) is reloaded and simulation continues in the same manner until an event $D_{i-1}$ occurs, the intermediate event $C_i$ occurs again, or a fixed number $R_{i-1}$ of simulation runs are performed.

If $N_A$ events $A$ occurred (in all repetitions) and $N$ events occurred together (in the first repetitions), an estimation of $P(A)$ can be determined by

$$\hat{P}(A) = \frac{N_A}{N \prod_{i=1}^{m} R_i}. \tag{3.32}$$

The variance of the result is formally given in relation to the covariance [226]. Unfortunately, determining the covariance is usually a difficult task.

## 3.2 Markov Chains

Simulation, as just described, provides a powerful method for performance evaluation. But it comes with a huge drawback: it often requires long run times until accurate results with high confidence levels are determined. Often, it is possible to approximatively model the real system using stochastic processes. If those stochastic processes belong to the class of Markov processes, mathematically treatable models result.

### 3.2.1 Markov Process

Markov processes are often proposed to model and evaluate parallel and distributed systems. A large theory is provided by [4, 19, 44, 53, 54, 89, 92, 95, 96, 199].

Markov processes are stochastic processes, characterized by a generalization of random variables and relationships between them. In a stochastic process, the random variables $Y_{t_n}$ belong to a family, with $t_n$ representing a time parameter, $0 < t_0 < t_1 < \dots$. A stochastic process is called a Markov process if $Y_{t_{n+1}}$ depends only on the previous value $Y_{t_n}$ and not on the earlier values $Y_{t_0}, Y_{t_1}, \dots, Y_{t_{n-1}}$:

$$P(Y_{t_{n+1}} \leq s_{n+1} \mid Y_{t_n} = s_n, Y_{t_{n-1}} = s_{n-1}, \dots, Y_{t_0} = s_0)$$
$$= P(Y_{t_{n+1}} \leq s_{n+1} \mid Y_{t_n} = s_n), \tag{3.33}$$

where $s_n$ represents the state of the system at time $t_n$. The Markov property (also called memoryless property) is often formulated as follows: the future of a process depends only on the present and is independent of the past. In consequence, the mean time between state changes (called mean sojourn time) is identical to the mean residual and the mean elapsed time [199].

If the system additionally shows a time-independent pattern of dynamic behavior,

$$P(Y_{t_{n+1}} \leq s_{n+1} \mid Y_{t_n} = s_n) = P(Y_{t_{n+1}-t_n} \leq s_{n+1} \mid Y_0 = s_n), \tag{3.34}$$

it is called time-homogeneous. Otherwise, it is said to be time-inhomogeneous.

If the state space representing the set of all states $s_n$ consists of discrete values (discrete state space), a Markov process is called Markov chain (Figs. 3.3(b) and 3.3(d)). In the following, only models dealing with a discrete state space are considered.

Time can also be distinguished to consist of continuous values (continuous time) or discrete values (discrete time), as shown in Fig. 3.3.

### 3.2.2 Discrete Time Markov Chain

This subsection deals with discrete values in time, before an extension to Markov chains with continuous time is presented in the next subsection.

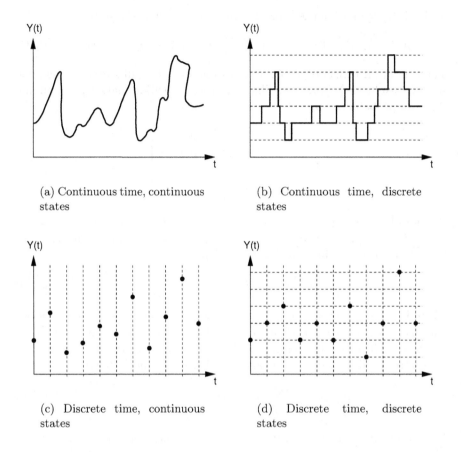

(a) Continuous time, continuous states

(b) Continuous time, discrete states

(c) Discrete time, continuous states

(d) Discrete time, discrete states

**Fig. 3.3.** Kinds of processes

## Definition

A Markov chain with a discrete time parameter $t_n$ is known as a discrete time Markov chain (DTMC). Equation (3.33) can be rewritten as

$$P(Y_{n+1} = s_{n+1} \mid Y_n = s_n, Y_{n-1} = s_{n-1}, \ldots, Y_0 = s_0)$$
$$= P(Y_{n+1} = s_{n+1} \mid Y_n = s_n), \qquad (3.35)$$

where $0, 1, \ldots, n, n+1, \ldots$ represent the points of observation. Due to the discrete time, the system changes its state step-by-step. Equation (3.35) yields the transition probability from state $s_n$ to state $s_{n+1}$. In the following, $s_n$ is abbreviated by $i$ and $s_{n+1}$ by $j$. Then, the transition probability $p_{ij}(n)$ changing from state $i$ to state $j$ in a single step at time $n$ is determined by

$$p_{ij}(n) = P(Y_{n+1} = s_{n+1} = j \mid Y_n = s_n = i) \qquad (3.36)$$

or, in the case of a homogeneous discrete time Markov chain, by

$$p_{ij} = p_{ij}(n) = P(Y_{n+1} = j \mid Y_n = i) = P(Y_1 = j \mid Y_0 = i). \qquad (3.37)$$

The sum of all transition probabilities from state $i$ to all states $j$ (including state $i$) results in $\sum_j p_{ij} = 1$. The stochastic transition matrix $\mathbf{P}$ accommodates all probabilities $p_{ij}$:

$$\mathbf{P} = [p_{ij}] = \begin{pmatrix} p_{00} & p_{01} & p_{02} & \cdots \\ p_{10} & p_{11} & p_{12} & \cdots \\ p_{20} & p_{21} & p_{22} & \cdots \\ \vdots & \vdots & \vdots & \end{pmatrix}. \qquad (3.38)$$

Besides the one-step transition probability, $n$-step transition probabilities $p_{ij}^{(n)}(k, l)$ are defined. They give the probability of changing from state $i$ at time $k$ to state $j$ at time $l$ in $n = l - k$ steps:

$$p_{ij}^{(n)}(k, l) = P(Y_l = j \mid Y_k = i), \qquad (3.39)$$

where $0 \le k \le l$. Due to the Markov property (independence assumption) and due to the constraint $\sum_j p_{ij}^{(n)}(k, l) = 1$, the $n$-step transition probability can be described by one-step probabilities if an intermediate state $h$ at time $m$ is introduced ($k < m < l$). $h$ can be each state of the state space:

$$p_{ij}^{(n)}(k, l) = \sum_h p_{ih}^{(m-k)}(k, m) \cdot p_{hj}^{(l-m)}(m, l). \qquad (3.40)$$

Recursively applying Eq. (3.40) leads back to the one-step transition probabilities. Equation (3.40) determines a system of equations called the Chapman-Kolmogorov equation. In the case of a homogeneous discrete time Markov chain, this equation becomes independent of $k$ and $l$,

$$p_{ij}^{(n)} = \sum_h p_{ih}^{(m)} \cdot p_{hj}^{(n-m)}, \qquad (3.41)$$

and leads, for $m = 1$, to

$$p_{ij}^{(n)} = \sum_h p_{ih}^{(1)} \cdot p_{hj}^{(n-1)}. \qquad (3.42)$$

Then, the matrix $\mathbf{P}^{(n)}$ of $n$-step transition probabilities is given by

$$\mathbf{P}^{(n)} = \mathbf{P}^{(1)}\mathbf{P}^{(n-1)} = \mathbf{P}\mathbf{P}^{(n-1)} = \mathbf{P}^n. \qquad (3.43)$$

This means that the $n$-step transition probability matrix is determined by multiplying the one-step transition probability matrix $n - 1$ times by itself.

Now, the transient state probabilities $\nu_i(n) = P(Y_n = i)$ that the Markov chain is in state $i$ at time $n$ can be calculated starting with the initial probabilities $\nu_i(0)$. If the vector $\boldsymbol{\nu}(n) = (\nu_0(n), \nu_1(n), \nu_2(n), \ldots)$ denotes all state probabilities, $\boldsymbol{\nu}(n)$ is given by

$$\boldsymbol{\nu}(n) = \boldsymbol{\nu}(0)\mathbf{P}^{(n)} = \boldsymbol{\nu}(0)\mathbf{P}^n = \boldsymbol{\nu}(n-1)\mathbf{P}. \tag{3.44}$$

The state probabilities at time $n$ represent an important measure to determine the performance of a system that is modeled by Markov chains. For instance, one of the states of a modeled network architecture can describe an entirely occupied buffer. Thus, sending a new message to this buffer would lead to a loss of the message. The model can be applied to investigate the occurrence of such states with transitions leading to message loss.

Many investigations examine the system in question particularly for their stationary behavior. In such a case, the system state does not change from time step to time step. In homogeneous DTMC, the stationary probability vector $\boldsymbol{\nu} = (\nu_0, \nu_1, \nu_2, \ldots)$ introduces such a system state: the transition probabilities $p_{ij}$ do not influence the state probabilities $\nu_j = \sum_i \nu_i p_{ij}$, for all $j$:

$$\boldsymbol{\nu} = \boldsymbol{\nu}\mathbf{P}, \tag{3.45}$$

while

$$\sum_i \nu_i = 1 \tag{3.46}$$

holds. If only a single stationary probability vector exists independently of the initial system state, this vector is called unique steady-state probability vector. If it exists, it can be derived from Eq. (3.44) via the limiting state probabilities $\tilde{\boldsymbol{\nu}}$:

$$\tilde{\boldsymbol{\nu}} = \lim_{n \to \infty} \boldsymbol{\nu}(n) = \lim_{n \to \infty} \boldsymbol{\nu}(n-1)\mathbf{P} = \tilde{\boldsymbol{\nu}}\mathbf{P}. \tag{3.47}$$

The steady state can usually be determined more simply by solving the system of equations as given above rather than by the time-dependent calculation of $\boldsymbol{\nu}(n)$. Many algorithms exist to solve a system of equations (if the Markov chain is of finite state). Nevertheless, difficulties arise if the system of equations becomes very large, or if the matrix $\mathbf{P}$ consists of a large number of "0" entries, or entries close to it. In the first case, the computer memory required to solve and store the system of equations may exceed the available memory. In the latter case, numerical rounding errors may cause inaccurate results and division-by-zero problems.

## Characteristics

Several characteristics of discrete time Markov chains are identified. They are briefly summarized here. Detailed definitions and characterizations can be found in [19].

If a state $j$ can occur by starting at state $i$ and performing a finite number of state transitions, the state $j$ is called reachable from state $i$. If all states are reachable from each other, the Markov chain is irreducible.

If no other state than itself can be reached from a state, the state is absorbing. In such a case, the Markov chain cannot be irreducible.

The states to which the Markov chain will return are said to be recurrent states. All other states are transient states. The period $D_i$ of a recurrent state $i$ is defined as the greatest common divisor of all $n$ that fulfills the condition $p_{ii}^{(n)} > 0$. If $D_i > 0$, it is a periodic state; otherwise, it is aperiodic. If an aperiodic state exists and the Markov chain is irreducible, then the Markov chain itself is aperiodic. If all states are additionally of finite mean recurrence time, the Markov chain is called ergodic.

Due to the Markov property, the sojourn time of a homogeneous DTMC is geometrically distributed: the probability that state $i$ will exist at time $k$ is given by

$$P(R_i = k) = (1 - p_{ii}) \cdot p_{ii}^{k-1}, \tag{3.48}$$

where $R_i$ represents the sojourn time of state $i$. Because of the geometric distribution, the mean sojourn time is

$$E(R_i) = \frac{1}{1 - p_{ii}}. \tag{3.49}$$

### 3.2.3 Continuous Time Markov Chain

Continuous time Markov chains (CTMCs) differ from discrete time Markov chains by the time at which state transitions may occur. In the case of CTMCs, transitions can take place at an arbitrary time, in contrast to DTMCs, where state transitions happen only at discrete points in time.

**Definition**

A Markov chain with a continuous time parameter $t_n \in \mathbb{R}_0^+$ is known as a continuous time Markov chain. Due to the discrete state space, Eq. (3.33) changes to

$$P(Y_{t_{n+1}} = s_{n+1} \mid Y_{t_n} = s_n, Y_{t_{n-1}} = s_{n-1}, \ldots, Y_{t_0} = s_0)$$
$$= P(Y_{t_{n+1}} = s_{n+1} \mid Y_{t_n} = s_n), \tag{3.50}$$

defining the Markov property in the continuous time case. As a result, the state sojourn time is exponentially distributed, because that is the only continuous memoryless distribution (see Sect. 2.2.2).

The transition probability $p_{ij}(t_a, t_b)$ of changing from state $i$ to state $j$ during the time interval $[t_a, t_b)$, with $t_a \leq t_b$, is given by

$$p_{ij}(t_a, t_b) = P(Y_{t_b} = s_b = j \mid Y_{t_a} = s_a = i), \tag{3.51}$$

or, in the case of a homogeneous continuous time Markov chain with $t = t_b - t_a$,

$$p_{ij}(t) = p_{ij}(0, t) = P(Y_{t_a+t} = j \mid Y_{t_a} = i) = P(Y_t = j \mid Y_0 = i). \quad (3.52)$$

If the probability of holding state $i$ at time $t_a$ is denoted by $\pi_i(t_a)$,

$$\pi_j(t_b) = \sum_i p_{ij}(t_a, t_b)\pi_i(t_a) \quad (3.53)$$

describes the probability of being in state $j$ at time $t_b$. All state probabilities can be included in a vector $\boldsymbol{\pi}(t) = (\pi_0(t), \pi_1(t), \pi_2(t), \dots)$ leading to

$$\boldsymbol{\pi}(t_b) = \boldsymbol{\pi}(t_a) \cdot \mathbf{P}(t_a, t_b), \quad (3.54)$$

with the stochastic transition matrix $\mathbf{P}(t_a, t_b) = [p_{ij}(t_a, t_b)]$. In the case of a homogeneous CTMC, Eq. (3.54) simplifies to

$$\boldsymbol{\pi}(t) = \boldsymbol{\pi}(0) \cdot \mathbf{P}(0, t) = \boldsymbol{\pi}(0) \cdot \mathbf{P}(t). \quad (3.55)$$

The Chapman-Kolmogorov equations of the continuous time Markov chain are defined by

$$p_{ij}(t_a, t_b) = \sum_k p_{ik}(t_a, t_c)p_{kj}(t_c, t_b) \quad (3.56)$$

with $t_a \leq t_c \leq t_b$. However, in contrast to the DTMC, a differential system of equations must be established to determine the steady-state probabilities. Thus, transition rates $q_{ij}(t)$ are needed. They describe the transition rate form state $i$ to state $j$ at time $t$:

$$q_{ij}(t) = \lim_{\Delta t \to 0} \frac{p_{ij}(t, t + \Delta t)}{\Delta t} \qquad \text{if } i \neq j, \quad (3.57)$$

$$q_{ii}(t) = \lim_{\Delta t \to 0} \frac{p_{ii}(t, t + \Delta t) - 1}{\Delta t}. \quad (3.58)$$

Applying these equations to Eq. (3.56) and with $t_c \to t_b$, the Kolmogorov forward equation is

$$\frac{\partial p_{ij}(t_a, t_b)}{\partial t_b} = \sum_k p_{ik}(t_a, t_b)q_{kj}(t_b). \quad (3.59)$$

The Kolmogorov backward equation can be established by applying $t_c \to t_a$:

$$\frac{\partial p_{ij}(t_a, t_b)}{\partial t_a} = \sum_k p_{kj}(t_a, t_b)q_{ik}(t_a). \quad (3.60)$$

The time-homogenous versions result in

$$\frac{dp_{ij}(t)}{dt} = \sum_k p_{ik}(t)q_{kj} \quad \text{and} \quad (3.61)$$

$$\frac{dp_{ij}(t)}{dt} = \sum_k p_{kj}(t)q_{ik}. \quad (3.62)$$

The differential equation for the unconditional state probabilities $\frac{d\pi_j(t_b)}{dt_b}$ obtained from Eq. (3.53) can be converted now by using the Kolmogorov forward and backward equations:

$$\frac{d\pi_j(t_b)}{dt_b} = \frac{\partial \sum_i p_{ij}(t_a, t_b)\pi_i(t_a)}{\partial t_b}$$

$$= \sum_k q_{kj}(t_b)\pi_k(t_b). \qquad (3.63)$$

In the time-homogenous case, Eq. (3.63) simplifies to

$$\frac{d\pi_j(t)}{dt} = \sum_k q_{kj}\pi_k(t). \qquad (3.64)$$

A matrix form of this equation can also be given as

$$\dot{\boldsymbol{\pi}}(t) = \frac{d\boldsymbol{\pi}(t)}{dt} = \boldsymbol{\pi}(t)\mathbf{Q}, \qquad (3.65)$$

where $\mathbf{Q}$ denotes the infinitesimal generator matrix defined by the transition rates from state $k$ to state $j$ (with $k \neq j$): $\mathbf{Q} = [q_{kj}]$. The diagonal elements $q_{kk}$ are determined such that the sum of each row of $\mathbf{Q}$ vanishes: $q_{kk} = -\sum_{j|j\neq k} q_{kj}$.

To determine the performance of a system, the steady state of the system is often investigated. The steady-state probability vector, also called the equilibrium probability vector, gives this information. If the steady state exists,

$$\lim_{t\to\infty} \frac{d\boldsymbol{\pi}(t)}{dt} = 0 \qquad (3.66)$$

holds because the steady-state probabilities are independent of time. In consequence, Eq. (3.64) leads to

$$0 = \sum_k q_{kj}\pi_k, \qquad (3.67)$$

for all $j$, resulting in a system of linear equations:

$$0 = \boldsymbol{\pi}\mathbf{Q}. \qquad (3.68)$$

**Characteristics**

Many solutions of the system of equations Eq. (3.68) exist, for instance, the trivial solution of all $\pi_i = 0$. A unique solution can be obtained via an additional condition. Then, the homogenous continuous time Markov chain is said to be ergodic. Such an additional condition is achieved by the fact that the sum of all steady-state probabilities must add up to 1:

$$\sum_i \pi_i = 1. \tag{3.69}$$

A homogenous CTMC can be ergodic only if it is irreducible. This means that each state $i$ is reachable from every other state $j$.

In contrast to discrete time Markov chains, continuous time Markov chains are never periodic. As a result, the mean recurrence time is finite for all states.

The state sojourn times $R_i$ of the homogenous CTMC must fulfill the memoryless property. This means that they are exponentially distributed with mean value

$$E(R_i) = -\frac{1}{q_{ii}}. \tag{3.70}$$

The residual state holding time equals the state sojourn time in distribution and mean. It describes the time until the next state change takes place.

### 3.2.4 Solution Methods

After having derived a Markov chain that models the system in question, a mathematical solution method must be applied to solve the Markov chain. As seen in the previous section, a system of equations represents the Markov chain, and must be solved. The steady-state solution and the transient solution can be identified. The algorithms of both solution methods are briefly described in the following. A detailed discussion of them is given in [19, 189]. Ergodicity of the Markov chains is assumed.

### Steady-state Solution

To determine the steady-state probability vector, either Eq. (3.45) or (3.68) must be solved in the case of a discrete time Markov chain or a continuous time Markov chain, respectively. If Eq. (3.45) is changed to $\mathbf{0} = \boldsymbol{\nu}(\mathbf{P} - \mathbf{I})$, where $\mathbf{I}$ represents the identity matrix, then both cases must deal with the solution of a linear system of equations,

$$\mathbf{0} = \mathbf{xA}. \tag{3.71}$$

The matrix $\mathbf{A}$ is of rank $n - 1$ if the Markov chain consists of $n$ states. This means that one of the equations is redundant. It can be replaced by the additional condition of Eq. (3.46) or Eq. (3.69), respectively. Then, commonly used solution methods of linear systems can be applied.

Some Markov chains allow symbolic solutions. Such a Markov chain is, for instance, a birth-death process. In a birth-death process, state transitions only occur between neighboring states. Thus, matrix $\mathbf{A}$ mainly consists of zero entries. Equation (3.71) can easily be solved by choosing one of the state probabilities and successively substituting all others in the right-hand sides of the equations.

Markov chains of general structures can be solved by numerical methods. One of them is called Gaussian elimination. The algorithm transfers the system of equations Eq. (3.71) into a triangular structure. This is done by solving the last (i.e., $n$-th) equation for $x_{n-1}$, and replacing $x_{n-1}$ in all other equations. Then, $x_{n-2}$ is similarly replaced by solving the $(n-1)$-th equation, and so on until $x_0$ is obtained. Then, successively substituting the state probabilities leads to the solution.

The Grassmann algorithm varies from the Gaussian elimination. It avoids subtraction to keep numerical results free of rounding errors in the case of nearly equal numbers. Nevertheless, it also yields a triangular matrix, as with the Gaussian elimination.

Besides direct methods like Grassmann's algorithm and Gaussian elimination, iterative methods also introduce a way to solve the system of equations Eq. (3.71). Iterative methods do not change the matrix $\mathbf{A}$. Thus, efficient methods to store this matrix can be applied. Systems of larger state space can be stored and solved, compared to direct methods. On the other hand, the iteration algorithm must ensure that it converges. Unfortunately, it is usually difficult to prove convergence.

Steady-state investigation is mostly performed by applying fixed point iteration [149]. Fixed point iteration solves equations of the form

$$\mathbf{x} = f(\mathbf{x}). \tag{3.72}$$

First, $\mathbf{x}$ is initialized with a value $\mathbf{x}^{(0)}$. Then, the new value of $\mathbf{x}$ is calculated in each iteration step $\ell$ from its previous value:

$$\mathbf{x}^{(\ell)} = f(\mathbf{x}^{(\ell-1)}). \tag{3.73}$$

The iteration is stopped if $\mathbf{x}^{(\ell)}$ is close to the exact solution $\mathbf{x}$ given by a tolerance level $\epsilon$. Because the exact solution is unknown, the accuracy is usually estimated by comparing the variance of the results $\mathbf{x}^{(\ell)}$ of several successive iteration steps.

Many methods of fixed point iteration exist. The power method determines the state probabilities by initializing them and then iteratively calculating

$$\boldsymbol{\nu}^{(\ell)} = \boldsymbol{\nu}^{(\ell-1)}\mathbf{P} \tag{3.74}$$

in the case of discrete time Markov chains or

$$\boldsymbol{\pi}^{(\ell)} = \boldsymbol{\pi}^{(\ell-1)}(\mathbf{Q} + \mathbf{I}) \tag{3.75}$$

in the case of continuous time Markov chains. If the convergence criterion $\epsilon$ is reached, the iteration is stopped.

Jacobi's method starts with a system of equations

$$\mathbf{b} = \mathbf{x}\mathbf{A}. \tag{3.76}$$

After initializing $\mathbf{x}$, an iteration step is given by

$$\mathbf{x}^{(\ell)} = \left(\mathbf{b} - \mathbf{x}^{(\ell-1)}(\mathbf{U} + \mathbf{L})\right)\mathbf{D}^{-1}, \tag{3.77}$$

where $\mathbf{D}$ represents the diagonal matrix of $\mathbf{A}$, $\mathbf{U}$ the upper triangular matrix, and $\mathbf{L}$ the lower triangular matrix: $\mathbf{A} = \mathbf{D} + \mathbf{U} + \mathbf{L}$.

A variation of this algorithm is given by the Gauss-Seidel method. All $x_i$ of Eq. (3.77) are sequentially determined instead of in parallel, as in Jacobi's method. An acceleration is achieved because already newly calculated values of state probabilities are considered while computing the other ones:

$$\mathbf{x}^{(\ell)} = \left(\mathbf{b} - \mathbf{x}^{(\ell)}\mathbf{U} - \mathbf{x}^{(\ell-1)}\mathbf{L}\right)\mathbf{D}^{-1}, \tag{3.78}$$

where $\mathbf{x}^{(\ell)}$ at the right-hand side of the equation represents the already calculated values of the current iteration step $\ell$.

**Transient Solution**

If the system performance at a particular time $t$ is of interest, a different solution method must be applied than the steady-state solution. Furthermore, there is no common method for discrete time Markov chains and continuous time Markov chains.

In the case of continuous time Markov chains, Eq. (3.65) builds the basis to be solved. Two kinds of measures can be determined by this equation: instantaneous measures, which give performance information for a point in time, and cumulative measures, which give performance information for a period of time. The latter change Eq. (3.65). If

$$\mathbf{L}(t) = \int_0^t \boldsymbol{\pi}(\tau)d\tau \tag{3.79}$$

denotes the overall amount of time spent in each state during the given time interval $[0, t)$, the differential equation

$$\frac{d\mathbf{L}(t)}{dt} = \mathbf{L}(t)\mathbf{Q} + \boldsymbol{\pi}(0), \tag{3.80}$$

with $\mathbf{L}(0) = \mathbf{0}$ must be transiently solved.

Symbolic transient solutions of continuous time Markov chains exist if the system is very simple, e.g., a pure birth process with constant birth rate $\lambda$. This results in the generator matrix

$$\mathbf{Q} = \begin{pmatrix} -\lambda & \lambda & 0 & 0 & \cdots \\ 0 & -\lambda & \lambda & 0 & \cdots \\ 0 & 0 & -\lambda & \lambda & \cdots \\ \vdots & \vdots & \vdots & \vdots & \end{pmatrix}. \tag{3.81}$$

This system of linear differential equations can successively be solved (see, for instance, [19]). The state probabilities are obtained as

$$\pi_k(t) = \frac{(\lambda t)^k}{k!} e^{-\lambda t} \tag{3.82}$$

for all states $k$.

More complex systems can only be determined by numerical methods. The transient uniformization (also called Jensen's method) represents such a method. Instantaneous measures can be calculated, as well as cumulative measures. Equation (3.65) is solved by embedding a discrete time Markov chain into a CTMC. The CTMC state probabilities are expressed as a power series of the transition probabilities of the DTMC [164].

In addition to uniformization, standard techniques to solve ordinary differential equations can be used. The time interval of interest is divided into several parts. Then, the solution is determined by considering the state probabilities at the discrete interval limit times. Explicit and implicit solutions exist [164].

**Decomposition**

Besides the methods to achieve accurate solutions as presented in previous sections, approximation methods can also be applied. Decomposition belongs to this group. Courtois [41, 42] invented this method for calculating the steady-state performance. It divides the model into submodels that can be investigated separately. Then, the results of the submodels are combined to obtain the global results of the model.

In the optimal case, the submodels are completely independent, and the results of the submodels represent the global results. But unfortunately, most models cannot be divided into completely independent submodels. Thus, nearly-independent submodels with tightly coupled structures should be found. In this case, the model itself is said to be nearly completely decomposable.

For instance, a system with several states offers an example if one (or more) of the states is very rarely visited. Due to the infrequent transitions to this state, they can be neglected, and this state and the rest of the state space can be assumed to be nearly-independent submodels. They are separated and independently solved.

Dividing a model into submodels leads to new generator matrices $\mathbf{Q}_{IJ}$ in the case of a CTMC or to new transition probability matrices $\mathbf{P}_{IJ}$ in the case of a DTMC as submatrices of the global one. For instance, if $\mathbf{P}$ is given by

$$\mathbf{P} = \begin{pmatrix} p_{00} & p_{01} & p_{02} & p_{03} & p_{04} & p_{05} & p_{06} & \cdots \\ p_{10} & p_{11} & p_{12} & p_{13} & p_{14} & p_{15} & p_{16} & \cdots \\ p_{20} & p_{21} & p_{22} & p_{23} & p_{24} & p_{25} & p_{26} & \cdots \\ p_{30} & p_{31} & p_{32} & p_{33} & p_{34} & p_{35} & p_{36} & \cdots \\ p_{40} & p_{41} & p_{42} & p_{43} & p_{44} & p_{45} & p_{46} & \cdots \\ p_{50} & p_{51} & p_{52} & p_{53} & p_{54} & p_{55} & p_{56} & \cdots \\ p_{60} & p_{61} & p_{62} & p_{63} & p_{64} & p_{65} & p_{66} & \cdots \\ \vdots & \vdots & \vdots & \vdots & \vdots & \vdots & \vdots \end{pmatrix}, \tag{3.83}$$

then the transition matrices of the submodels may be denoted as

$$\mathbf{P}_{00} = \begin{pmatrix} p_{00} & p_{01} \\ p_{10} & p_{11} \end{pmatrix}, \tag{3.84}$$

$$\mathbf{P}_{11} = \begin{pmatrix} p_{22} & p_{23} & p_{24} \\ p_{32} & p_{33} & p_{34} \\ p_{42} & p_{43} & p_{44} \end{pmatrix}, \text{ and} \tag{3.85}$$

$$\mathbf{P}_{22} = \begin{pmatrix} p_{55} & p_{56} & \cdots \\ p_{65} & p_{66} & \cdots \\ \vdots & \vdots \end{pmatrix} \tag{3.86}$$

if $M = 3$ nearly-independent submodels are generated with the first submodel consisting of states 0 and 1, the second submodel consisting of state 2, 3, and 4, and the third submodel consisting of the remaining states. The remaining transition probabilities $p_{ij}$ build the additional transition matrices $\mathbf{P}_{IJ}$ with $I \neq J$. This means that the matrix $\mathbf{P}$ is in general given by

$$\mathbf{P} = \begin{pmatrix} \mathbf{P}_{00} & \mathbf{P}_{01} & \cdots & \mathbf{P}_{0(M-1)} \\ \mathbf{P}_{10} & \mathbf{P}_{11} & \cdots & \mathbf{P}_{1(M-1)} \\ \vdots & \vdots & & \vdots \\ \mathbf{P}_{(M-1)0} & \mathbf{P}_{(M-1)1} & \cdots & \mathbf{P}_{(M-1)(M-1)} \end{pmatrix} \tag{3.87}$$

$$= \mathbf{A} + \mathbf{B}, \tag{3.88}$$

where $\mathbf{A}$ consists of all $\mathbf{P}_{ii}$ and $\mathbf{B}$ of the remaining submatrices:

$$\mathbf{A} = \begin{pmatrix} \mathbf{P}_{00} & & & & \\ & \mathbf{P}_{11} & & \mathbf{0} & \\ & & \mathbf{P}_{22} & & \\ & \mathbf{0} & & \ddots & \\ & & & & \mathbf{P}_{(M-1)(M-1)} \end{pmatrix}, \tag{3.89}$$

$$\mathbf{B} = \begin{pmatrix} \mathbf{0} & \mathbf{P}_{01} & \cdots & \mathbf{P}_{0(M-1)} \\ \mathbf{P}_{10} & \mathbf{0} & \cdots & \mathbf{P}_{1(M-1)} \\ \vdots & \vdots & & \vdots \\ \mathbf{P}_{(M-1)0} & \mathbf{P}_{(M-1)1} & \cdots & \mathbf{0} \end{pmatrix}. \tag{3.90}$$

Up to now, the transition matrices $\mathbf{P}_{II}$ are not stochastic matrices, because the elements of each row may not sum up to 1. A matrix $\mathbf{X}$ must be introduced so that stochastic matrices $\mathbf{P}^*$ and $\mathbf{P}^*_{II}$ result:

$$\mathbf{P}^* = (\mathbf{A} + \mathbf{X}) = \begin{pmatrix} \mathbf{P}^*_{00} & & & \\ & \mathbf{P}^*_{11} & & \mathbf{0} \\ & & \mathbf{P}^*_{22} & \\ & \mathbf{0} & & \cdots \\ & & & & \mathbf{P}^*_{(M-1)(M-1)} \end{pmatrix}. \tag{3.91}$$

The matrix $\mathbf{X}$ should be defined such that matrices $\mathbf{P}^*_{II}$ become ergodic. Equations (3.88) and (3.91) lead to

$$\mathbf{P} = (\mathbf{A} + \mathbf{X}) + (\mathbf{B} - \mathbf{X}) \tag{3.92}$$
$$= \mathbf{P}^* + \mathbf{C}. \tag{3.93}$$

Now, all submodels $I$ can separately be solved using

$$\boldsymbol{\nu}^*_I \mathbf{P}^*_{II} = \boldsymbol{\nu}^*_I \tag{3.94}$$

to obtain their corresponding state probabilities $\boldsymbol{\nu}^*_I$. With these probabilities, a transition probability matrix

$$\boldsymbol{\Gamma} = [\Gamma_{IJ}] = \begin{pmatrix} \Gamma_{00} & \Gamma_{01} & \cdots & \Gamma_{0(M-1)} \\ \Gamma_{10} & \Gamma_{11} & \cdots & \Gamma_{1(M-1)} \\ \vdots & \vdots & & \vdots \\ \Gamma_{(M-1)0} & \Gamma_{(M-1)1} & \cdots & \Gamma_{(M-1)(M-1)} \end{pmatrix} \tag{3.95}$$

can be determined. The states $\Gamma_{IJ}$ are called macro states and calculated by

$$\Gamma_{IJ} = \sum_i \left( \nu^*_{I_i} \sum_j p_{I J_{ij}} \right), \tag{3.96}$$

where $i$ and $j$ represent the row and column of the denoted matrix or vector, respectively. With $\boldsymbol{\Gamma}$, the macro steady-state probabilities can be obtained by

$$\boldsymbol{\gamma}\boldsymbol{\Gamma} = \boldsymbol{\gamma}, \tag{3.97}$$

with $\sum \gamma_I = 1$. Then, an approximation of the steady-state probabilities $\boldsymbol{\nu}$ of the initial model can be found by

$$\nu_{I_i} \approx \gamma_I \nu^*_{I_i}. \tag{3.98}$$

The error of this approximation is investigated in [43].

Besides this method of Courtois, another decomposition method called Takahashi's method exists [193]. Again, a decomposition into macro states must be performed. But the macro states of Takahashi's method are established differently than that of Courtois. In contrast to Courtois' method, no nearly-independent submodels must be found. This is because the macro state probabilities and the original state probabilities are iteratively determined. Unfortunately, there is no generally applicable rule for clustering the system. But the efficiency and convergence of this method depend very much on good clustering. A detailed description of the method can be found in [19, 193].

## 3.3 Petri Nets

Modeling a parallel or distributed system with Markov chains may lead to a very large system of equations. For the modeler, it becomes difficult to handle all the equations and to avoid errors. Therefore, many high-level modeling languages (e.g., queuing networks, fault-trees, production rule systems, etc. [79]) have been introduced to keep the model development as simple as possible. Many of those high-level languages map their resulting model onto a lower-level description. Petri nets [160] belong to this group of high-level modeling languages. If they fulfill particular requirements, they can be mapped onto a continuous time Markov chain.

### 3.3.1 Basic Petri Nets

Petri nets were first introduced as a modeling tool in automata theory. Later, many extensions were proposed to improve their modeling power.

#### Definition

Petri nets [36, 63, 75, 86, 118, 128, 144, 165] are a bipartite directed graph consisting of two kinds of nodes and two kinds of arcs. The nodes are denoted as places, $p_i$, and transitions, $t_i$. The arcs are either input arcs or output arcs. A place is connected to a transition via an input arc. A connection from a transition to a place is established via an output arc. Arcs between the same kinds of nodes are not allowed. Places are usually graphically represented by circles, and transitions are depicted by rectangles or bars.

A Petri net furthermore consists of tokens represented by dots within places. A particular distribution of tokens in a set $P$ of places is called a marking $M \in \mathbb{N}^{|P|}$. It determines the number of tokens in each place and represents a particular state of the Petri net. Figure 3.4 presents a small example of a Petri net with two places, two transitions (each with an input arc and an output arc), and a token in place $p_1$.

If there is at least one token in all places connected to a particular transition by an input arc, the transition is said to be enabled. Then, an atomic action called firing may take place. The firing of a transition results in the

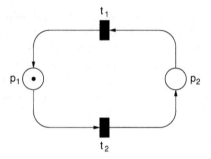

**Fig. 3.4.** Example of a Petri net

removal of a token from each place connected to the transition by an input arc and in the addition of a token to each place connected to the transition by an output arc. If more than one transition is enabled, the firing must be sequentialized. No parallel firing is allowed. Petri net semantics does not define which transition fires first.

Due to the removal and addition of tokens, the firing of a transition changes the marking of the Petri net. Thus, firing leads to a state change of the model. The behavior of the system in question can be described by a sequence of firings. The initial state of the system is indicated by the initial marking $M_0$ of the Petri net model. A marking that can be reached from the initial marking by a sequence of firings is called reachable. The reachability set consists of all such markings.

Taking into consideration previous explanations, a Petri net definition is as follows: a Petri net is described by a 5-tuple

$$\mathbf{PN} = \{P, T, I, O, M_0\}, \tag{3.99}$$

with

$$
\begin{aligned}
P &= \{p_1, \cdots, p_{|P|}\} &&: \text{set of places} \\
T &= \{t_1, \cdots, t_{|T|}\} &&: \text{set of transitions} \\
I &\in \{0,1\}^{|P \times T|} &&: \text{matrix of input arcs} \\
O &\in \{0,1\}^{|P \times T|} &&: \text{matrix of output arcs} \\
M_0 &= \{m_{01}, \cdots, m_{0|P|}\} &&: \text{initial marking}
\end{aligned}
$$

and $m_{0i}$ representing the number of initial tokens in place $p_i$. An input arc from place $p_i$ to transition $t_j$ exists if and only if $I_{p_i,t_j} = 1$. Accordingly, an output arc from transition $t_j$ to place $p_i$ exists if and only if $O_{p_i,t_j} = 1$.

### Extensions

The basic definition of Petri nets as presented above has been extended in two ways: to improve modeling power and to improve modeling convenience.

Improving modeling convenience only changes the way a model represents the system under consideration. Such extensions can be unfolded to achieve again a basic Petri net that conforms to the definition.

Introducing an arc multiplicity is one of the extensions. Arc multiplicities deal with multiple input arcs or output arcs between two nodes of a Petri net. Instead of drawing $k$ input (output) arcs connecting place $p_i$ and transition $t_j$, only a single input (output) arc is depicted. The multiplicity $k$ is denoted on the arc. The entries of $I$ and $O$ represent the multiplicities, and thus are natural numbers.

In case of multiplicities, a transition is only enabled if the number of tokens in a place corresponds at least to the multiplicity of the input arc connecting the place and the transition.

An extension to improve modeling power are inhibitor arcs. An inhibitor arc connects a place $p_i$ and a transition $t_j$. Its graphical representation is a line starting at the place and ending with a small circle at the transition. If place $p_i$ holds at least one token, transition $t_j$ is disabled independently of the state of the other places. Arc multiplicities may also be allowed. Then, transition $t_j$ is disabled if there are at least as many tokens in place $p_i$ as the multiplicity denotes.

Inhibitor arcs help check whether a place holds a given number of tokens. Petri nets of the basic definition are not able to do so. With inhibitor arcs, their modeling power is increased.

Modeling power is also increased by defining transition priorities. If transitions are connected via input arcs to the same place, and this place does not hold a sufficient number of tokens to allow all the transitions to fire, transition priorities regulate which transitions are preferred. In such a case, the transition of the highest priority only will be enabled (if its other input arcs are connected to places that also hold a sufficient number of tokens). In the graphical representation, priorities are denoted as integers at the transition.

Another extension of the modeling power is given by guards. As inhibitor arcs, they disable transitions depending on the marking of the Petri net. But they are more general: a transition may only be enabled if the corresponding guard is satisfied. The guard may consist of any marking-dependent condition.

Considering previous explanations, an extended Petri net definition is as follows: an extended Petri net is described by an 8-tuple

$$\mathbf{PN}_{\mathrm{ext}} = \{P, T, \mathcal{P}, I, O, H, G, M_0\}, \tag{3.100}$$

with

$$P = \{p_1, \cdots, p_{|P|}\} \qquad : \text{set of places}$$
$$T = \{t_1, \cdots, t_{|T|}\} \qquad : \text{set of transitions}$$
$$\mathcal{P} \in \mathbb{N}^{|T|} \qquad\qquad : \text{vector of transition priorities}$$
$$I \in \mathbb{N}^{|P \times T|} \qquad\quad : \text{matrix of input arcs and multiplicities}$$
$$O \in \mathbb{N}^{|P \times T|} \qquad\quad : \text{matrix of output arcs and multiplicities}$$
$$H \in \mathbb{N}^{|P \times T|} \qquad\quad : \text{matrix of inhibitor arcs and multiplicities}$$
$$G \in \{\text{true}, \text{false}\}^{|T|} \quad : \text{vector of guards}$$
$$M_0 = \{m_{01}, \cdots, m_{0|P|}\} \quad : \text{initial marking}$$

where $G$ is a marking dependent function.

### 3.3.2 Stochastic Petri Nets

The previous section defined basic Petri nets. Stochastic Petri nets (SPNs) extend this definition so that stochastic events and timed events can also be handled, further improving the modeling power [141, 146].

### Timed Transitions

Timing is introduced by adding a new parameter to transitions. This parameter, called firing time, defines the time interval from the enabling of a transition until it fires (if there is no influence from any other transition that fires).

If multiple timed transitions are enabled, an execution policy must determine which transition is allowed to fire. For instance, the preselection policy chooses one of the enabled transitions at random. Firing times are not considered. A policy more often applied is called race policy, and will be used in the following chapters. In race policy, firing times are considered and the transition with the smallest firing time fires first.

To determine what happens to the remaining transitions that did not fire, a memory policy must be introduced. For instance, the remaining transitions may again restart their firing times, and time already spent in an enabled state is lost. This policy is called resampling policy. In contrast, the enabling memory policy conserves the time that has already been spent in an enabled state. After the firing of a transition, the firing times of the remaining transitions continue to elapse if they are still enabled. If they are no longer enabled, the firing time already spent is lost. A third policy is called age policy. In this policy, firing times already spent are never lost.

Different kinds of memory policies may be applied in a single stochastic Petri net. Each transition can follow its given memory policy independently of the other transitions to represent its desired behavior.

If each place connected to a transition via an input arc is covered by multiple tokens, the transition can act either as a single server transition or as an infinite server transition. A single server transition processes the tokens in sequence. This means that the firing time related to a particular token starts to elapse after the firing process related to the previous one concludes. In contrast, an infinite server transition processes all tokens in parallel. It behaves similarly to an infinite number of parallel single server transitions.

Firing times can be of any probability distribution to incorporate stochastic events. If transitions show geometric or exponential distribution, the stochastic Petri net may be transformed to a discrete time or continuous time Markov chain. Then, the solution methods of DTMCs or CTMCs can be applied to determine the performance of the modeled system (see Sect. 3.3.2). If only transitions with exponential distributed firing times (including the special case of transitions with zero firing times) are used, such nets are called generalized stochastic Petri nets.

## Generalized Stochastic Petri Nets

Generalized stochastic Petri nets (GSPNs) [129] are Petri nets that only consist of transitions with exponential distributed firing times or with zero firing times. Transitions with zero firing time are said to be immediate transitions. Transitions with exponentially distributed firing times are graphically represented as white rectangular boxes, while immediate transitions are black segments.

Particularly in the case of immediate transitions, several transitions may be enabled in parallel. Due to zero firing times, firing has to take place simultaneously. Then, firing probabilities help determine the order of firing. Such firing probabilities are introduced by associating weights $w_i$ to transitions. The normalized weight $w_i/(\sum_j w_j)$ of a transition $i$ (normalized to all enabled transitions $j$) gives its firing probability.

A marking in which at least one immediate transition is enabled is called vanishing marking because it immediately changes due to zero firing time. All other markings are called tangible markings.

The exponentially distributed firing time of a transition is characterized by its firing rate $\lambda_i$. The firing rate may be marking dependent. A firing rate $\lambda_i \rightarrow \infty$ denotes a zero firing time, and thus represents an immediate transition in a GSPN definition.

Previous explanations lead to a definition of generalized stochastic Petri nets described by a 10-tuple:

$$\mathbf{GSPN} = \{P, T, \mathcal{P}, I, O, H, G, M_0, \Lambda, W\}, \tag{3.101}$$

with

$$P = \{p_1, \cdots, p_{|P|}\} \qquad \text{: set of places}$$
$$T = \{t_1, \cdots, t_{|T|}\} \qquad \text{: set of transitions}$$
$$\mathcal{P} \in \mathbb{N}^{|T|} \qquad \text{: vector of transition priorities}$$
$$I \in \mathbb{N}^{|P \times T|} \qquad \text{: matrix of input arcs and multiplicities}$$
$$O \in \mathbb{N}^{|P \times T|} \qquad \text{: matrix of output arcs and multiplicities}$$
$$H \in \mathbb{N}^{|P \times T|} \qquad \text{: matrix of inhibitor arcs and multiplicities}$$
$$G \in \{\text{true}, \text{false}\}^{|T|} \qquad \text{: vector of guards}$$
$$M_0 = \{m_{01}, \cdots, m_{0|P|}\} \qquad \text{: initial marking}$$
$$\Lambda \in (\mathbb{R}^+ \cup \{\infty\})^{|T|} \qquad \text{: vector of transition firing rates}$$
$$W \in \mathbb{R}^{+|T|} \qquad \text{: vector of transition weights.}$$

Extensions of generalized stochastic Petri nets also exist. Often, systems show deterministic behavior concerning some events. To adapt the modeling power of GSPNs to such behavior, transitions with deterministic firing times are added to the given definition [130]. The resulting Petri net that consists of transitions of immediate, exponentially distributed, and deterministic firing times is called a deterministic and stochastic Petri net (DSPN). Transitions with deterministic firing times are mutually exclusive and have a preemption policy. They are graphically represented as black rectangular boxes.

## Solution Methods

Using stochastic Petri nets as a modeling technique, performance measures of the system under consideration can be obtained by various observations. For instance, a particular marking may reveal a particular event to be investigated. Then, the probability that this marking occurs can be determined. Furthermore, the number of tokens in a place may give some performance information. The frequency of the firing of a transition could also be of interest. The delay of a token that traverses a subset of the Petri net is also often investigated for performance evaluation.

After establishing a GSPN or DSPN model of the system in question, analytical or simulation methods can be applied to determine the performance measures. For instance, discrete event simulation, as presented in Sect. 3.1 provides a solution method.

A mathematical solution of Petri nets is available for GSPNs based on the extended reachability graph. It includes the reachability set and some stochastic information at the arcs. Timed transitions generate firing rates at the corresponding arcs, and immediate transitions generate probabilities. The graph can be established by starting with the initial marking $M_0$ as root. Next, all possible firing patterns result in the next possible markings. Then again, all possible firing patterns are treated, and so on.

The extended reachability graph describes a semi-Markov process. It can be transformed into a continuous time Markov chain by the elimination of all vanishing markings: probabilities/rates of an arc leading to the vanishing marking and exiting from it are simply multiplied, resulting in a new arc that skips this vanishing marking [8, 19]. The resulting graph is called reduced reachability graph. Usually, the reduced reachability graph is generated on the fly instead of the memory-consuming extended reachability graph.

The reduced reachability graph can be solved by applying steady-state analysis or transient analysis, as presented in Sect. 3.2.

If multiple transitions with deterministic firing times exist, the Petri net cannot generally be mapped onto a Markov chain. Only DSPNs, which consist of deterministic transitions with mutually exclusive firing times and preemption policy can be solved by using the Markov chain formalism. They can be replaced by an embedded Markov chain [62, 63, 90], where the system is investigated at time instants at which it is memoryless. Such instants are called regeneration points. They are time instants when either transitions with an exponentially distributed firing time only are enabled or when a deterministic transition is enabled and fires. A discrete time Markov chain at the regeneration points can be established. A continuous time Markov chain describes the exponentially distributed state transitions while the deterministic transition is enabled. The resulting embedded Markov chain can be solved and the investigated performance measures determined [63].

# 4

# Model Engineering

Previous chapters outlined common modeling techniques as well as characteristics of network architectures for parallel and distributed systems. This chapter aims to combine both. The topic of network architecture modeling is exhaustively discussed by considering the experience acquired while performing the research that has been taken as a basis for this book. Model development is addressed first. Reducing the model size and the handling of the evaluation process are also dealt with. Finally, model validation often points out the limits of a model. This chapter gives some guidelines

1. for keeping the model to a size tractable by a computer system and
2. for keeping the required time (to establish and solve the model) as small as possible.

The first item refers to the often observed effect that an established model exceeds available computer memory due to its size.

The second item addresses the constraints of available computing power and man power: the time interval between the modeling request and the reception of the final results, referred to as modeling time, should ideally last only an instant. But it may in practice last minutes, hours, or even days. The goal of reducing modeling time leads to the two tasks of decreasing the model development time (mainly in the case of mathematical modeling methods) and the model evaluation time (also called computation time) for performance evaluation (mainly in the case of simulation).

## 4.1 Model Development

The most important issues in model development are model size, modeling time, and the accuracy of the model's results. All three issues heavily depend on each other: the more accurate the results are to be, the more detailed the model must be, something that usually increases the modeling and evaluation times.

Many things influence model size, modeling time, and accuracy. Some of them may be influenced and adapted during the modeling process. But some of them are out of the control of the model developer due to system constraints and modeling goals. Modeling aspects include

- modeling technique,
- modeling granularity,
- logical system complexity,
- system complexity in time, and
- performance measures.

The modeling technique affects model size, modeling time, and accuracy, as well as the proceeding model development processes. Before exhaustively discussing this, other aspects are briefly addressed.

The modeling granularity describes how detailed and exact the system in question is represented by the model. It is obvious that a more detailed and exact model leads to a larger model size, because either a higher number of system states are distinguished, or a higher number of transitions between states. As a result, modeling time (development time as well as computation time) is increased. On the other hand, higher accuracy is usually achieved. Fewer system dependences are neglected, driving the model behavior closer to the real-world system behavior.

Logical system complexity gives information about the number of system components and their logical dependence on each other. The more the number of system components and logical dependences, the larger the model size. The reason is similar to that of the previous case of the modeling granularity: the number of system states to be distinguished and the number of transitions increase with logical system complexity. Of course, logical system complexity strongly depends on the point of view from which the system is observed. For instance, the complexity of a distributed or parallel system hardware differs if it is observed at transistor level, compared to being observed at the logical gate level. Nevertheless, both levels achieve exact results, in contrast to low granularity, where dependences are neglected.

System complexity in time gives information about the number of system state changes per time interval. As with logical system complexity, system complexity in time strongly depends on the point of view the system being observed. For instance, if a clocked system is modeled, either all system state changes during a single clock cycle may be separately modeled, or, if a less detailed point of view is chosen, they are accumulated. As a result in the latter case, only system state at the end of the clock cycle must be considered: the number of system states to be distinguished and the number of transitions can be reduced, leading to a smaller model size.

Performance measures to be determined also influence model size and modeling time. The more detailed the performance measures required, the more detailed the model to be established, increasing model size and modeling time. For instance, delay times of messages sent between the nodes of a distributed system are requested. Determining the mean delay of all messages leads to a less complex model than determining the delay distribution, for which different delays have to be separately dealt with.

As already mentioned, the most important consideration concerning model size, modeling time, and accuracy is the kind of modeling technique chosen for performance evaluation. Selecting a particular modeling technique sets the modeling limits related to it as presented in Chap. 3. Often, mathematical modeling techniques turn out not to be appropriate due to their restrictions and model sizes.

If both simulation and mathematical modeling are feasible, then the optimal technique usually depends on the number of investigations [201] that are planned. Due to short model development times and long model computation times, simulation is the better choice if only one (or a few) investigation(s) must be performed. Mathematical modeling methods are a better choice if a large number of investigations are required: developing a mathematical model consumes much time, but evaluating this model usually leads to the result very fast.

Table 4.1 gives an example. While writing this book, many ways to model multistage interconnection networks (MINs) were developed and investigated (see Chap. 6 for details). Their modeling times are compared in Table 4.1.

**Table 4.1.** Comparison of the modeling methods

| Modeling method | Development time | Computation time |
|---|---|---|
| Petri net | 100 person hours | > 2 weeks |
| *MINSimulate* | 400 person hours | 4 hours |
| Markov chains | 1500 person hours | < 1 second |

The Petri net description [212] of a MIN is set up using the graphical toolkit *TimeNET* [65]. Due to the huge state space resulting from the Petri net description, no analytical solution is feasible. Simulation must be performed to achieve performance results of the MIN. The simulation tool *MINSimulate* [208], which is based on the programming language C++, was specifically developed to model and evaluate MINs. Finally, the mathematical model is based on Markov chains. Note that this mathematical model includes many neglects and simplifications to keep the state space small enough to be tractable by a computer system. Section 4.2 points out all neglects and simplifications and explains them.

Table 4.1 first compares the model development times. The estimated model development times include the validation and error detection times of the models. There is a large difference in the development times. A fast method in development and model validation is the Petri net description. The main reason is the high level of abstraction, leading to a simple and compact graphical representation. It allows a proper model survey. The token game supports easy model validation. The time consumed for the C++ simulation program development is higher than for Petri nets especially because model validation is more complicated. The validation requires a step-by-step simulation. The development time of the mathematical model is much higher than in all other methods. The reason is the low abstraction level development based on Markov chains. The state transition probabilities are time dependent, which leads to complex equations.

The evaluation times of the methods are as different as the development times, but in reverse order. The mathematical modeling method is the fastest method for results. All other models use simulation, which enlarges the evaluation time. The C++ simulation run time is lower than the Petri net run time because this simulator profits from being written specifically for model multistage interconnection networks. Table 4.1 shows the evaluation times (computation times) on a Linux PC consisting of a 1,200 MHz processor. The throughput, delay times, and queue lengths at all stages of a 64×64 network consisting of 2×2 switching elements are investigated. The MIN has a buffer length of 1 at each stage, an input rate of 1.0, and a multicast input traffic pattern called traffic$_{eqpr}$ [211].

Because of the high simulation times in the case of large network sizes, a mathematical model is preferable if many networks with different parameters have to be evaluated. Petri net descriptions are an acceptable method if the examination of small networks is sufficient or if a single network has to be evaluated.

If no idea at all is available about the complexity of the network of a parallel and distributed system, the best way to start is with a modeling technique of least development time, such as Petri nets. Then, a model is established in a short time, and the first practical experience in modeling this particular system is gained. If it turns out that the model computation time is too high, not much time has been wasted in starting with this simple model development technique. Moreover, the first experience is usually very helpful to understand how to model the basic behavior of the system. This knowledge accelerates the development of models using other modeling techniques like Markov chains.

The flow chart of Fig. 4.1 gives a rough guideline for determining the modeling technique. Of course, there is usually no clear answer to most questions in the decision boxes due to the imprecise questions, and maybe due to the poorly known system. But the developer's tendency (how he is inclined to answer questions) also helps find a feasible modeling technique with the given guidelines.

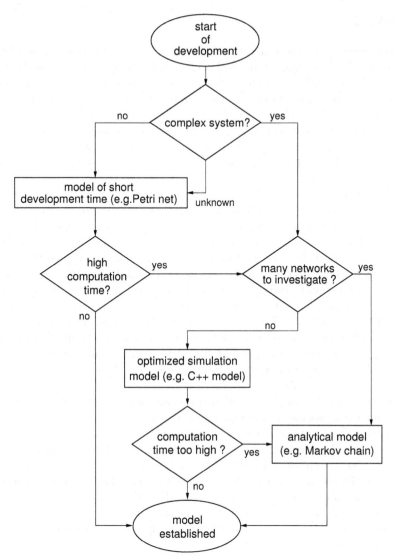

**Fig. 4.1.** Model development guideline

## 4.1.1 Simulation Model

If simulation is chosen as the modeling technique, the drawback of high evaluation times surfaces. There are several reasons evaluation times could increase more than necessary.

**Overhead**

One of these reasons was mentioned above. Table 4.1 includes two simulation models: a Petri net simulation and a C++ simulation. While evaluating the same MIN, very different simulation run times are needed. The Petri net model takes much more time to obtain the results due to Petri net formalism, and that overhead has to be dealt with.

Figure 6.1 of Chap. 6 gives an example. It shows the Petri net description of an 8×8 MIN consisting of 2×2 SEs. Without explaining the detailed behavior here (it will be explained in Chap. 6), the complex Petri net structure is nevertheless obvious.

Using a modeling method like Petri nets may lead to a model that deals not only with the functional behavior of the system in question, but additionally deals with all intern relations and dependences of the method itself, such as priorities and weights. For instance, in the above example there are many immediate transitions that feature equal priority and equal weight (e.g., all Store transitions). With regard to the functional behavior, the order of firing does not matter once the transitions allowed to fire are fixed. Only the system states before and after the firing of the Store transitions are relevant. Intermediate states are not of interest. But using Petri nets and their definitions results in the choosing of a single transition at random to fire. Afterwards, the state of the Petri net is revaluated. Again, many transitions are discovered to have equal priority and weight, and one of these is chosen and fires, revaluation starts again, and so on.

Such a model behavior is called overhead: details are considered during the simulation process that are irrelevant concerning the functional behavior. Simulating these details increases the computation time.

Modeling the functional behavior without any overhead confirms this: Table 4.1 references a second simulation method of multistage interconnection networks. The toolkit *MINSimulate* [208] simulates networks of the same kind as those previously modeled by Petri nets. This time, the toolkit was specifically developed for modeling multistage interconnection networks. Simulation is performed by C++ code. A detailed description of *MINSimulate* can be found in Sect. 6.2.

Even though some overhead exists due to the object-oriented design, *MINSimulate* shows much less overhead than exists in the Petri net case due to its particular design. As a result, model evaluation time is reduced to a fraction of Petri net run time. On the other hand, developing toolkits like *MINSimulate* consumes much more time than taking advantage of a commonly available toolkit like *TimeNET*, which supports establishing complex system models.

**Data Representation**

Besides an efficient simulation algorithm with as little overhead as possible, an efficient simulation data representation also accelerates simulation.

*MINSimulate* serves as an example: packets that enter the multistage interconnection network are stored at a dedicated computer memory location. This includes the headers of the packets and all the payload. During the packet's movement from buffer to buffer through the MIN, only a pointer to the packet is moved from data structure to data structure, which represent the buffers. If all packet data were to be moved, this process would be much more time consuming; simulation run time would increase.

Such an efficient data representation also shows another direction for accelerating simulation: data that are only locally available in a parallel and distributed system in the real world can be made globally available in simulation to decrease simulation run time. For instance, in the real world, multistage interconnection networks consist of signaling lines between neighboring stages. The lines signal to preceding stages whether the packet destination buffers are available so packets may be forwarded to them. If cut-through switching is applied, a packet may cross many stages during a clock cycle. This means that signaling information must also cross several stages to be available to intermediate switching elements, and finally at the packets stage.

Simulating this local signaling consumes much more time than smartly accessing all buffers globally to discover which ones are empty and directly forwarding the packet to the last buffer of the determined cut-through path of the current clock cycle. Applying this local information globally in simulation does not change the overall behavior of how far packets move compared to the real world. Advantage is only taken of the fact that in a simulation program, all status information can be globally addressed, while in real world, this information is only locally available in a parallel and distributed system due to hardware constraints.

For instance, *MINSimulate* profits from globally addressed information that is only locally available in the real world.

## Confidence

If simulation is applied and the model includes stochastic events, the confidence level of the achieved results must be determined. For instance, the confidence level of *MINSimulate*'s simulation results is observed by the toolkit *Akaroa* [155]. The simulation is stopped when previously established termination criteria are met.

The chosen confidence level dramatically influences the model evaluation time. The more accurate the results are required to be, the higher the evaluation time. For instance, the C++ model evaluation time of Table 4.1 emerges by choosing a confidence level of 95% and an estimated precision of 2%. If the confidence level is changed to 98%, the model evaluation time increases by more than 30 minutes. If the confidence level is kept at 95% and the estimated precision is changed to 1%, model evaluation time is dramatically increased to about 11 hours.

This effect is particularly observed when simulating rare events. Loss rates of networks are an example. Loss rates of about $10^{-9}$ lead to an average number of a billion packets crossing the network until a packet is lost: high evaluation times arise. If a very narrow confidence interval with a high precision is additionally chosen, many packet loss events must occur until the simulation termination criteria are met, increasing the model evaluation time to intolerable values.

As a result, confidence level and estimated precision should be carefully chosen. The required accuracy of the simulation results should be considered to determine the lowest reasonable confidence level and the highest reasonable estimated precision value, avoiding unnecessary model evaluation times.

With regard to confidence, another issue may increase model evaluation time: the random number generator, which is needed if stochastic events are modeled. As already mentioned in Sect. 3.1.1, random number generators generate numbers in cycles. This means that after generating a certain quantity of numbers, a previously generated number is generated again. The same sequence of numbers starts again due to the generation algorithm.

If the cycle is very short and the root of the random numbers is badly chosen, the observed measure may start to oscillate with equal period around the average value. If this oscillation is of high amplitude, the desired confidence level may never be reached. Therefore (and to avoid wrong results), it is important to provide simulation with random number generators of large cycles. *MINSimulate* incorporates the random number generator of *Akaroa* instead of the ANSI C rand() function. *Akaroa* offers periods between $2^{185}$ and $2^{377}$ [109, 154], while ANSI C rand() offers periods of only $2^{31}$.

### Parallelism and Data Recycling

A further reduction of model evaluation time can be achieved by distributed simulation on several computer systems or processors that are connected via a network. A simulation distributing the entire model to each computer or processor is based on multiple replications in parallel (MRIP), as presented in Sect. 3.1.2. If the model in question is simulated on $i$ computer systems in parallel, model evaluation time is accelerated by slightly less than $i$ due to two effects: first, additional communication between the $i$ computer systems is needed, consuming time; second, the initial transient phase must be performed by all computers. Thus, this time is not divided by $i$.

But often, communication time and simulation time until the steady state is reached can be neglected. For instance, *MINSimulate* in combination with *Akaroa* offers MRIP. Profiting from this scheme halves simulation time of the previously presented example of a 64×64 MIN to about two hours if executed in parallel on two computers instead of about four hours for a stand-alone simulation.

Due to the multiple replications independently running on different processors, fault tolerance is included in the simulation. If a replication on a processor fails, the remaining replications still deliver their results, and the simulation still comes to a successful end as soon as termination criteria are met. Additionally, a restart of the failed replication can be performed.

Another type of parallel measure achievement emerges from determining multiple measures in parallel. If multiple measures of a system must be determined, it is obvious to integrate them into a single system model, so starting the simulation once will provide all results instead of starting it multiple times, once for each measure. This is particularly relevant if terminating simulation results are of interest. If results for several points in time are desired, all observed parameters have to be determined for each time step in question. In discrete time systems, these time steps are usually the discrete time steps of the system. For instance if the system is clocked, a clock cycle equals a time step. In continuous time systems, an appropriate time step must be chosen related to the desired investigation of the system.

Determining temporal parameters in separate simulations for each time step is not efficient. A faster method observes all parameters of all time steps in a single simulation. The only drawback of such a method is the number of measures. For instance, if 100 parameters are observed at 1,000 time steps, 100,000 measures must be handled by the simulator.

### 4.1.2 Mathematical Model

If a mathematical technique like Markov chains or Petri nets is chosen for modeling, granularity and complexity become very important issues. That is because mathematical models allocate considerable memory on a computer system. Usually, a mathematical model leads to a system of equations describing the states and the state transitions of the system in question. If the state space is too large, it exceeds the memory of computer systems, and the system of equations cannot be solved.

An example is given by the Petri net model of Fig. 6.1. As mentioned in Chap. 3, Petri nets help mathematically model systems. Petri nets can be mapped onto the underlying Markov chain leading to the related system of equations. In the case of the model of Fig. 6.1, the maximum size of a MIN modeled in such a way must not be larger than $2\times2$. The state space of larger MINs significantly increases and results in state space explosion: they are not tractable by an ordinary computer system.

Reduction of the modeling granularity to reduce the state space would lead to a very inaccurate model, and is therefore not feasible. Nevertheless, due to the Petri net definitions, the translation of the Petri net to the Markov chain introduces many states that are not relevant for the general behavior. These states may be superposed to a single state. This concept is called lumping.

The previous example shows that establishing a minimum-sized model is one of the main tasks in mathematical modeling. Usually, high level description techniques like Petri nets are not able to optimize model size. Therefore, achieving small model size is closely related to development at the low description level of Markov chains, for instance. But even that may result in models too large to be handled by computer systems. In such a case, model size must be further reduced by neglecting system dependences and by decreasing the state space. The model will be simplified.

## 4.2 Complexity Reduction

This section discusses model simplifications to reduce model complexity. The aim is either to speed up model simulation run time or to allow mathematical modeling of systems whose mathematical model was too large and formerly not tractable.

Simplifying models such that their complexity is very low is an easy task. Simply changing the modeling granularity to a very low value would do it. But in this case, the accuracy of the performance evaluated by the model would drop very much, and the results would be useless. Therefore, the challenge is to find simplifications that only slightly change the accuracy. Usually, these simplifications keep the modeling granularity. In the following, the general approaches that turned out to be efficient are summarized.

### 4.2.1 Simulation

As previously mentioned, high evaluation times are the main drawback of simulations. Reducing model complexity usually helps accelerate simulation. Two different ways to reduce model complexity with only slightly decreasing accuracy are presented here: neglecting dependences and combining a sequence of events into a single event.

#### Neglects of Dependences

Neglecting dependences reduces the number of state transitions and states. In other words, simulation runs pass fewer transitions and states, and therefore simulation is accelerated.

On the other hand, dependence neglects usually influence the behavior of the model. This means that the model behavior differs from the real-world system behavior. In consequence, model results also differ from system results. This can be avoided only if dependences that only slightly influence model behavior are neglected (these dependences are called loose dependences). Then, model results will differ only slightly from real-world system results.

Applying dependence neglects to models therefore gives rise to the task of finding loose dependences in the system. There is no general way to do this. Usually, a system developer should know the system enough to identify obviously loose dependences. But there may be others that are not immediately visible, or ones where it is unclear whether the dependence is loose or strong.

If no clear statement about the kind of dependence is possible, it must be investigated by comparing the model results by including the dependence and then excluding it. Of course, such a comparison only makes sense if the model is built to allow more than one investigation. Then, the results of both models can be compared for a few carefully selected investigations. If the less complex model turns out to be nearly as accurate as the more complex one, it can be assumed that such a behavior is true for all investigations. The less complex model can be exclusively applied to the remaining investigations: simulation run time is reduced.

An example demonstrates dependence neglects [220]. In a distributed system, messages are sent from node to node through the interconnection network. If packet switching is applied, these messages are divided into packets (see Fig. 2.1). All packets belonging to the same message are destined to the same node and enter the network in sequence. This is also true for the packets belonging to the next message, intended for another destination node, and so on. In the real distributed system, sequences of packets with the same destination enter the network. The upper MIN input of Fig. 4.2 illustrates such sequences. The first message is destined to node 3. It consists of four packets. The next message consists of two packets and is destined to node 1.

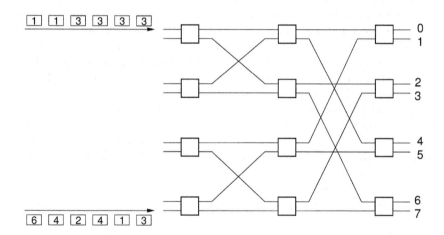

**Fig. 4.2.** Packet sequences

Neglecting these dependences in the network model means that all packets entering the network are assumed to be independent. Nodes are modeled as packet sources that generate packets with destinations independently of previously generated packets (lower MIN input of Fig. 4.2). Packet generation becomes less complex in simulation. Furthermore, performance measures vary less, and narrow confidence intervals are reached faster. In mathematical models, this particular dependence neglect raises additional issues, described in Sect. 4.2.2.

To give an example, two 64×64 MINs are compared. They are identical except for their nodes. The nodes of one network always offer sequences of 10 packets with the same destination to the MIN, while those of the other offer independent packets. Performance measures show that the throughput, delay times, and queue lengths of the MIN with the packet sequences as load are slightly smaller than those of the other. Simulation time (using the tool *MINSimulate*) is more than doubled.

In real networks of distributed systems, the dependence between packets is given not only by their destination. The packet density (number of packets per time interval) often depends on former densities, also in long time range. An example is multifractal traffic, which is often observed in networks. *MINSimulate* provides independent packet traffic as well as any kind of destination-dependent and time-dependent traffic.

### Combining Events

Combining a sequence of events into a single event may also decrease model evaluation time without decreasing accuracy. Each event that occurs and must be dealt with consumes computation time. In other words, reducing the events also reduces computation time. But the same problems occur as in case of neglected dependences: combining events may result in model behavior different from real-world system behavior, particularly if some events in the sequence are neglected. In consequence, results achieved from the model also differ from system results. The model becomes inaccurate.

To avoid inaccuracy, events have to be determined that can be combined without changing model behavior. A multistage interconnection network may serve as an example. As described in Sect. 2.3.10, such networks are internally clocked to achieve synchronously operating switches. But clocking only affects the network itself. The nodes of a distributed system, which are connected to the MIN, may operate asynchronously. Therefore, nodes offer packets to be transferred by the network at an arbitrary moment to the network. But the MIN only checks at the beginning of every clock cycle whether packets are waiting for transfer. In consequence, if packets are buffered at the nodes, all events related to the offering of a new packet to a network input can be combined. Only the number of packets waiting at the beginning of a clock cycle are of interest. Figure 4.3 illustrates asynchronous packet generation of nodes (top) and the combination of those events (bottom) into a single event at the beginning of a clock cycle (represented by a dashed line).

### 4.2.2 Mathematical Model

The main drawback of mathematical models is their usually large state space due to model complexity. Therefore, mathematical modeling aims to reduce

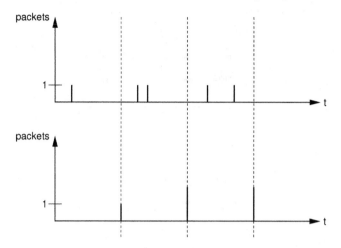

**Fig. 4.3.** Combining events

model complexity, as in the case of simulation. Ideas emerging from simulation can also be applied, but the features of mathematical models call for additional methods.

## Decomposition

One of the most effective methods to reduce model complexity is given by dividing the system to be modeled into multiple independent subsystems. Each of the subsystems results in a much smaller model than a model of the complete system. Subsystem models can be solved independently of each other to characterize the complete system. This approach is called decomposition (see Sect. 3.2.4).

An example is given by a four-node parallel system where each of two crossbars connects two nodes, and no cross-connection between the crossbars exists (Fig. 4.4). The system can be divided into two independent subsystems. Subsystem I consists only of nodes N1 and N2, while Subsystem II consists only of nodes N3 and N4. Both subsystems can be solved independently of each other.

Unfortunately, most systems cannot be divided into independent subsystems. In that case, some systems offer the facility to divide them into near-independent subsystems. If dependences can be ordered, their ordering determines the order of the solution. Figure 4.5 shows a parallel system with near-independent subsystems. If the link from the lower crossbar output of Subsystem I represents a unidirectional link to the crossbar of Subsystem II (and no backward signaling exists), an ordered solution of the subsystems solves the whole system. First, Subsystem I is solved; the lower crossbar out-

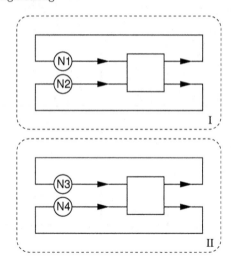

**Fig. 4.4.** Decomposition of independent parallel subsystems

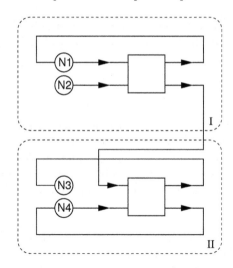

**Fig. 4.5.** Decomposition of near-independent parallel subsystems

put gives the offered load to the upper crossbar input of Subsystem II. Solving Subsystem II concludes the analysis of the system.

If the link between Subsystem I and II, for instance, also includes a backward signaling from II to I, and the backpressure algorithm is applied, then no ordering of the subsystems is possible, because dependences are cyclic. Two solution methods can be applied: an approximation, as described in Sect. 3.2.4, or an iterative method.

The iterative method solves the system using fixed point iteration. During each iteration step, the state of each subsystem is separately determined by considering the subsystem state of the previous iteration step. At every $i$-th iteration step, the influence of the other subsystems is incorporated. The less dependent the subsystems, the larger the $i$ that can be chosen. On the other hand, if $i$ is too large relative to the dependence of the subsystems, the iteration may oscillate and never reach a fixed point.

## Symmetries

Networks, particularly those of parallel systems, often belong to a class that profits in two ways from decomposition. First, dividing the networks into near-independent subsystems results in smaller models of subsystems that are tractable by a computer system. Additionally, this splitting into subsystems often results in another advantage: all subsystems are similar. In consequence, iteratively solving a single subsystem while it cyclically depends on itself saves computation time, compared to iteratively solving all subsystems.

Usually, similar subsystems emerge if the network is of regular (symmetric) structure. Not only the architecture, but network traffic must also be symmetric (uniform), i.e., similar in all subsystems. Then, substitution by a single subsystem is feasible.

Multistage interconnection networks are again an example. Figure 4.6(a) illustrates a buffered $8 \times 8$ MIN consisting of $2 \times 2$ switching elements. The network is of a regular structure, called modified data manipulator. It is assumed that the traffic offered to the network results in uniform traffic. This means that equal load is offered on average at all network inputs, and all outputs are destinations of the packets with equal probability. Then, all SE rows of the MIN show on average identical behavior and identical state probabilities as, for instance, the buffer queue length probabilities. The network model complexity can be reduced by simply modeling a single row of the MIN. It represents a subsystem and is given in Fig. 4.6(b). This subsystem cyclically depends on itself. For instance, for performance evaluation of the MIN for a given input traffic, the kind and amount of traffic at both inputs at the SE at stage 0 can immediately be determined. But Fig. 4.6(b) also shows that performance evaluation needs the kind and amount of traffic entering the model at the lower input of the SE at stage 1. This traffic comes from an output of an SE at stage 0 of another row. Due to symmetry, this SE sends on average the same kind and amount of traffic as the SE at stage 0 given in Fig. 4.6(b). In consequence, the average throughput at the SE output at stage 0 determines the offered load at the lower SE input at stage 1. On the other hand, the buffer queue length probabilities of the lower SE input at stage 1 determine the average amount of accepted packets and, therefore, the throughput the SE output at stage 0 can achieve. A cyclic dependence exists.

This scheme of reducing the MIN to a single row of SEs can be continued. An even more reduced model emerges from reducing each SE of the row to

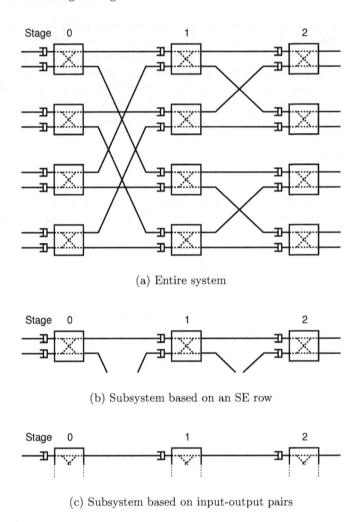

(a) Entire system

(b) Subsystem based on an SE row

(c) Subsystem based on input-output pairs

**Fig. 4.6.** 8×8 multistage interconnection network

a single input-output pair. Due to the SE symmetry and the uniform traffic, all SE inputs as well as all SE outputs show equal behavior: a single input-output pair models the entire SE. Thus, a row of input-output pairs, each of them representing an SE, is sufficient to model the entire MIN. In Fig. 4.6(b), this would be realized by horizontally cutting the figure in the middle of the SEs and considering only the upper part (Fig. 4.6(c)). The state of a MIN stage can now be represented and modeled by the state of a single buffer of the related SE input. An example is given in Chap. 6. There, a model representing a MIN is exhaustively discussed. The model is based on discrete time Markov chains and takes advantage of MIN symmetries.

## Asymmetry

Even if the system is not symmetric, the method above can be applied in particular cases. The idea is based on the superposition of symmetric system parts.

Often, asymmetric systems can be divided into parts where each of them shows symmetric structure. Multistage interconnection networks serve again as an example. Figure 4.7 illustrates a MIN of symmetric architecture but

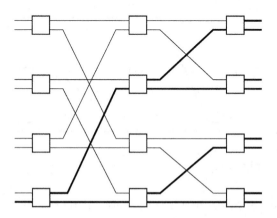

**Fig. 4.7.** MIN loaded with asymmetric traffic

asymmetric (non-uniform) network traffic: a very low load is offered to all network inputs except the last one. This input is fed by a high network load. The high load of one input spreads over the network and establishes a tree of high load in the network. As a result, the network traffic and, therefore, the buffer queue length probabilities are not equal in all SEs belonging to the same network stage.

The system can be modeled in a way similar to symmetric systems: system parts are modeled, which cyclically influence themselves. But, this time, a model not only of a single subsystem emerges from the system, but depending on the kind of asymmetry, multiple subsystems (model parts) are established. Dependences may exist between some of the subsystems and, cyclically, within a subsystem. A model reduction can be achieved by applying a single subsystem model part to all subsystems identical to it.

For instance, in Fig. 4.7, the part of the overall model that describes the SE of the first row at stage 0 also describes the SE of the third row at stage 0. Both SEs behave identically, and only a single model part describing one

of them is needed to establish a model of the whole system. In contrast, the
SE of the last row at stage 0 must present its own model part because its
behavior is different from that of above rows.

Using this scheme, only six model parts describing SEs emerge to establish
the whole system model instead of the 12 model parts if each SE is described
by its own model part. Similarities and resulting model parts are depicted in
Fig. 4.8. For instance, model part VI represents all four SEs of the last stage.

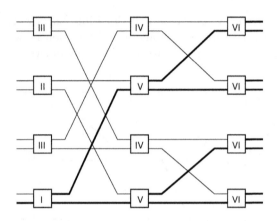

**Fig. 4.8.** Model parts of the asymmetric MIN

On the other hand, model part I is the only one representing a single SE.

### Multiple State Spaces

As already mentioned, very detailed models often result in huge state spaces
that cannot be handled by typical computer memory. Thus, those models
cannot mathematically be solved by computers.

If various different details of the system are relevant and should be modeled, a single state space representing all details may become too large. Establishing multiple models can solve the problem. Each of the models is described
by a state space representing one of the system details of interest. These multiple models are often closely related and their mathematical solution can
simultaneously be obtained from a single calculation.

A buffered network of a parallel or distributed system is an example: a
particular buffer of the network is investigated. The buffer is able to store a
maximum of 10 packets. By modeling the network, the aim is to achieve the
distribution of the queue lengths of the buffers. In addition, a distribution of
which packet holds the first buffer position is of interest. For instance, the
ratio between unicast packets and multicast packets may be examined.

The given problem can be solved by establishing a model with a single
state space. Because 10 different queue lengths can occur (if an empty buffer

is omitted), and because for each queue length there may be a unicast packet or multicast packet at the first position, 20 different states and their transitions need to be modeled.

If two models are established, two state spaces represent the system. One model investigates the queue length and thus consists of 10 states and their transitions. The other model examines the first buffer position resulting in two states and the corresponding transitions. That means if both models are simultaneously solved, only 12 states have to be handled instead of 20. The description of the system is of course not as detailed as before, but it may be sufficient for the desired investigation.

Representing a system by multiple models may induce multiple state spaces that are dependent on each other. In consequence, a simple product of state probabilities originating from different state spaces does not give the probability with which those states simultaneously occur.

The above example of two network models illustrates this problem. Let us assume that there are many more multicast packets in the first buffer position if the queue consists only of this position, and that there are fewer multicast packets if the buffer queue length is larger than a single packet. Thus, the probabilities of both states of the first buffer position only give the (weighted) *average* probabilities of *all* queue lengths. Simply multiplying the probability that the first buffer position is occupied by a multicast packet with the probability that the queue length is 1 does not result in the probability that the packet of a queue of length 1 is a multicast packet. The product would be too small because it would not consider that there are many more multicast packets if the queue consists only of one position.

## Discretization

In a discrete time system such as the MINs presented as clocked systems, state transition only occur at multiples of a time unit $\Delta t$. Markov chains only require the system to show the Markov property at discrete times. Only the state transition probabilities of two succeeding time units are needed. How much time the transitions consume is insignificant, as long as they finish in a time unit. The system model becomes less complex this way, and it is easier to deal with.

However, most real-world systems are continuous time systems. Nevertheless, many of them can be discretized by finding the smallest time unit that is relevant for their behavior. For instance, if a state variable of a system is to be controlled and the highest frequency of the system is known, then system time can be discretized to twice this frequency: Shannon's Theorem proves that a sample rate of twice the system frequency is sufficient to determine the system. But discretization may lead to a more detailed system description in terms of system states. A larger state space, as in the case of a continuous time model may result (as already concluded, the size of the state space is one of the most critical problems in mathematical modeling).

## 4.3 Automatic Model Generation

In previous sections, model development time and model evaluation time (computation time) were distinguished. Table 4.1 compares them for different modeling methods. It turns out that development of mathematical models like Markov chains consumes much more time than development of simulation models. The large development times arise due to the complicated and complex equation systems that have to be established for mathematical models.

If the equations that describe a system are investigated in detail, it turns out that, for many systems, not all the equations are completely different. Groups of equations exist in which the equations differ only slightly.

Interconnection networks of parallel and distributed systems are an example, for instance, if buffers are in the network. The state of each buffer can be described by the number of packets that stay in the buffer at a discrete point in time (e.g., at the beginning of a clock cycle). A state transition from a queue length of, for instance, four packets in a particular clock cycle to three packets in the succeeding one will take place if and only if a packet leaves the buffer and no new packet arrives. It is assumed that the buffer can only handle a single arrival and a single exit of a packet per clock cycle.

Now, consider the scenario, where a state transition from a queue length of five packets to four packets is investigated. Again, the same conditions hold: the transition takes place if and only if a packet leaves the buffer and no new packet arrives. This means that the equations modeling both state transitions belong to the same group. The conditions for the transition can be used as rules to set up the equations: in this example which, is quite simple, the transition probability is the product of the probability that a packet leaves the buffer and the probability that no new one arrives. Such rules can automatically be translated into equations, saving development time.

Deriving the rules for equation set-up based on the equation group emerges as the main issue for automatic model generation.

### 4.3.1 Rule Design

Before rules can be derived, the states of the system in question must be identified. To perform this task, the same considerations apply as in the case of manually setting up the equations.

For instance, the more the number of states defined, the more detailed the model. On the other hand, the more the number of existing states, the more complex the model, and the more the number of difficulties that arise in the mathematical solution of the equations. Complexity reduction, as described in Sect. 4.2, remains an important task. It must be emphasized that automatic model generation does not influence the model complexity compared to manual model set-up. If states are identical, then state transition equations are also identical. Automatic model generation only accelerates establishing a model.

After the states to be modeled have been identified, all existing state transitions must be derived. The next step deals with the determination of groups of similar state transitions and, thus, similar equations. These groups can be found by starting with a single equation. Designing rules to set up this single equation usually suggests which other state transitions also fit the rules.

Rules should be elaborated such that they clearly relate to a system characteristic. Then, corresponding rules can easily be located to adapt to new or changed system characteristics in order to investigate the new or changed system.

For instance, a distributed system is connected by a clocked and buffered network operating in store-and-forward switching mode. Rules are derived, describing, for example, the preconditions for the arrival of a new packet at a buffer. One of those rules states that at least one packet must stay in the preceding buffer in the previous clock cycle (Fig. 4.9). Otherwise, no packet would be available to be forwarded to the buffer described by the rule. Now,

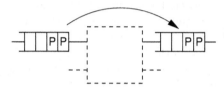

**Fig. 4.9.** Store-and-forward switching: only preceding buffer relevant

the switching mode is changed to cut-through switching. The rule above is clearly related to the switching mode and, thus, it is obvious that it must be changed. In the given example, it will be changed in such a way that an empty preceding buffer is also feasible if the predecessor of the preceding buffer includes at least one packet (Fig. 4.10), or the predecessor of the predecessor, and so on. In such cases, a packet "cuts" through all empty buffers until it

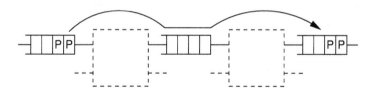

**Fig. 4.10.** Cut-through switching: preceding buffer may be empty

reaches the buffer in question.

Usually, the best way to elaborate the rules consists of investigating a small system. For instance, if a parallel system of a thousand nodes is to be investigated, and the system has regular structure, the first step would be to investigate a similar system of fewer nodes, e.g., 10 nodes. But the system must not be shrunk so much that it results in an extreme example with simpler equations and rules.

After the basic rules of a small system are found, they can be applied to larger systems. Larger systems show whether the rules are still valid or whether any add-ons are needed. On the other hand, it must be ensured that the rules also cover any extreme example.

Furthermore, any borderline cases must be considered while deriving the rules of a group. In a group, there are often equations that fit into the concept of the group but differ in a particular feature or show additional constraints. An example is equations describing transitions at the border of a regular structure, compared to equations for transitions in the middle of the regular structure.

Again, a distributed system connected by a clocked and buffered network operating in store-and-forward switching mode is an example. One of the rules describing the preconditions for the arrival of a new packet at a buffer was outlined above: at least one packet must stay in the preceding buffer at the previous clock cycle (Fig. 4.9). If such a packet is offered to the buffer in question, it will be stored in the buffer. This is true for all states of the queue length and, therefore, can be considered in all equations describing all states. But there is an exception: if the buffer in question is completely occupied by packets, it cannot accept the offered packet. In contrast to all other states, a packet must leave the buffer at the same time to accept a new one. In other words, the group of equations that describe the transition between two buffer queue lengths consists of similar equations, but one of them deals with an additional constraint. The rules must take this into account.

Similar considerations are well known from algebraic specification methods; for instance, specifying the functional behavior of an ADT buffer (abstract data type buffer) while considering full and empty buffer states.

### 4.3.2 Generating Systems of Equations

The systems of equations are generated based on rules. Section 4.3.1 helps find such rules. After deriving the rules, they must be represented in an appropriate way for computer processing.

In many cases, simple if-then constructs in an ordinary programming language are sufficient and the most efficient way. Nevertheless, other approaches may also be a qualified choice depending on the kinds of rules. For instance, Sect. 6.4 demonstrates an automatically generated system of equations for multistage interconnection networks. There, the idea was to represent the rules using an expert system. However, it turned out that an expert system would complicate the description more than necessary and if-then constructs were sufficient. Nevertheless, expert systems are also the feasible representations.

Another consideration in the design of an equation generator is how to include model parameters. Model parameters are, for instance, the size (number of nodes) of a parallel and distributed system, as well as the functional behavior like the switching technique (store-and-forward or cut-through). Three different schemes to include model parameters can be distinguished:

- hard coding of parameter values into rules,
- input of parameter values during the generation of equations, and
- input of parameter values during the solution of the equations.

These schemes may also be combined if the model consists of multiple types of parameters.

The hard coding of parameter values into the rules is usually necessary if multiple rule sets exist. A set of rules depends on the parameter value. The rules generate a specific system of equations valid only for the particular parameter value. If the value changes, new equations have to be generated. Such a scheme is a good option if rules significantly differ for parameter values. For instance, the switching technique in networks significantly influences a part of the rules to establish a model (see Sect. 4.3.1).

Entering the parameter values during the run-time of the generation of equations means that the rules are able to handle all (legal) parameter values. The rules are universal. Nevertheless, once the parameters are passed, a specific system of equations is generated that is again valid only for the particular parameter values. If the values change, new equations have to be generated.

Passing the parameter values during the run-time of the equations' solution means that the parameters (and not particular values) are included in the system of equations. Universal rules generate universal equations, which can be solved for all (legal) parameter values. This is the best scheme if a series of parameter values is to be investigated: the solution can be started multiple times with different parameters without generating new equations for each run. Such a scheme is usually a good option if the parameter value is simply a number, like an integer. For instance, the size of a parallel system represents such a parameter, or the buffer sizes of its network (see Sect. 4.3.1).

All equations belonging to the same group are generated by repetitively applying the rules of this group to each equation. The rules of this group combined with the particular constraints of each equation lead to its final form. The queue length of a FIFO buffer gives an example (Fig. 4.11). All equations to determine $\pi_m$ are to be generated, where $\pi_m$ represents the probability that the buffer queue length equals $m$.

**Fig. 4.11.** Queue length probabilities

A loop passes through all queue lengths $m$, starting with $m = 1$ and ending with $m = m_{max}$, where $m_{max}$ denotes the buffer size. For each $m$, the corresponding equation of $\pi_m$ is derived by considering the group rules and the particular constraints. For instance, one of the group rules says that a queue length of $m$ is reached if there was a queue length of $m + 1$ before, no new packet arrived at the buffer, and one left it. But if $m = m_{max}$, which means a completely filled buffer, the state $m + 1$ does not exist and this particular rule does not apply; nevertheless, maybe all other rules still apply. The nonexistent state $m_{max} + 1$ is a constraint for this equation.

In the given example, the loop passes through all states of the subsystem. Of course, state transitions may also belong to a group of equations, and, then, another loop passes through all those transitions. Many more groups may exist; for instance, measures may belong to a group.

Before automatically generating the equations by rules, another design issue must be considered: an appropriate way to represent the equations must be found. It has to be kept in mind what the equations model. For instance, that a parallel or distributed system is to be modeled to determine its temporal or steady-state performance. In other words, by considering the models goal, a step-by-step solution of each equation, a steady-state solution of the system of equations, or possibly another kind of solution will lead to the desired result. Therefore, many tools are feasible for handling the system of equations, depending on what the model is intended for. The equations must be represented in such a way that they can be imported into the corresponding tool. Tools may be

- computer algebra systems and tools or
- ordinary programming languages.

Of course, they are not limited to the above list.

Computer algebra systems usually support symbolic and numerical solutions of systems of equations. A symbolic solution allows an investigation of the desired measure depending on all parameters influencing it. For instance, if the measure results in a product of a parameter and any term of the remaining parameters, the value of the measure increases if the value of the first parameter increases. Specifically, the measure increases proportionally to the first parameter.

Symbolic solutions work very well if equations are polynomial and the number of variables is low. But often, such constraints are exceeded, and a symbolic solution fails. Then, a numerical solution can usually still be found. The computer algebra system yields a numerical value for the desired measure. The performance of the parallel or distributed system is determined for a particular case, but if parameters change, no prediction of how performance changes is possible.

Most computer algebra systems support various import formats, such as all kinds of data files and programming languages (e.g., C and Fortran). The equations generated should apply to one of these formats.

The system of equations can also be solved by algorithms implemented in ordinary programming languages. Any computational overhead arising from computer algebra systems due to their universality can be avoided. The model developer adapts the solution algorithm to his or her particular model and includes the automatically generated equations. For instance, fixed point iteration could act as a solution method, and be implemented in the C++ programming language. This main program never changes, and includes the equations generated, which are also represented by C++ code. If generation rules change due to a changed system, only the newly generated equations are included in the unchanged main program. After recompiling it, a new investigation may start.

### 4.3.3 Generator Design

Figure 4.12 graphically outlines previous explanations on how to design a generator for automatic model development and on how to evaluate the model. The figure represents a guideline for what steps to perform to achieve automatic model generation. Of course, there are various kinds of systems, and various model types that can be applied. Therefore, this guideline must be adapted to the particular problem. Considering all hints given in Sects. 4.3.1 and 4.3.2 will help. A complete example of an automatic generation of equations is given in Chap. 6.

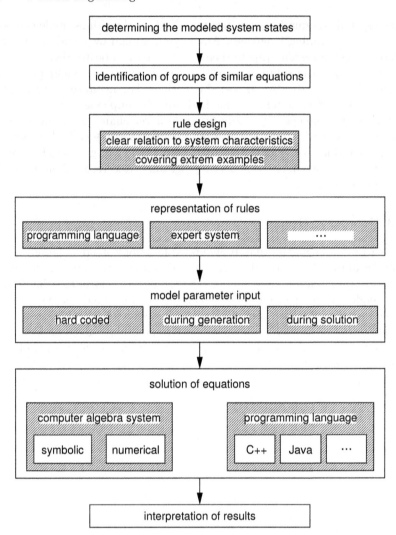

**Fig. 4.12.** Design of a generator and model evaluation

# 5

# Application: Cellular Network

Chapter 4 gave guidelines on how to model network architectures for parallel and distributed systems. Some ideas to reduce model complexity and accelerate model development were presented. This chapter gives an example on how to design a model for cellular networks using those ideas. The handoff in cellular networks will be of particular interest. The handoff process gives important information about the size of overlap area that must be chosen. But additional parameters, such as the velocity of the mobile nodes and the cell diameter, must also be considered.

The example investigates a concept for ubiquitous network access by mobile users dealing with real-time applications. They move in a cellular network and perform several handoffs. It is assumed that mobile users in particular use audio-based and video-based applications with specific quality of service (QoS) requirements. Support of real-time applications in wireless scenarios requires both fast handoffs and seamless QoS guarantees on the path of the mobile node through the network. A proposed solution to combine them for real-time support is the Ubiquitous Service Access Internet Architecture (USAIA) [183, 223]. USAIA provides hierarchical mobility management that interacts with the QoS mechanisms on three different network levels:

- The cellular level for handoffs between adjacent cells belonging to the same subnetwork. This level provides a fast handoff protocol and a scheme for resource reservation in advance to minimize the impacts of the movements of the mobile nodes to the QoS contract.
- The domain level for handoffs between different subnetworks. This level provides mobility management by means of either hierarchical foreign agents or Multiprotocol Label Switching (MPLS) [170]. Furthermore, QoS support is assumed to be provided by Differentiated Services and/or MPLS.
- The inter-networking level for handoffs between administrative domains. This level uses Mobile Internet Protocol (Mobile IP) [157] for mobility management.

For this investigation, the handoff is simplified, and the different protocols [188] are only marginally considered and not explained in detail here. The network performance will be mainly examined, depending on the network architecture and the required bandwidth of the mobile users.

## 5.1 USAIA Framework

USAIA is an all-IP framework for mobile access, with support for real-time traffic. It provides hierarchical mobility management that interacts with appropriate QoS mechanisms in different network areas. USAIA distinguishes the following three areas:

- the cellular level, including all cells of the wireless part of the access network belonging to the same IP subnetwork
- the domain level, including all IP subnetworks belonging to an administrative domain
- the global level, including all other administrative domains of the global Internet

Within the USAIA framework, base stations (BSs) are viewed as routers connecting the wireless cellular network to the wired part of the access network, which itself is connected to the Internet. Because the cellular level of USAIA is the subject of the performance evaluation of this study, it is explained in more detail in the following. All other aspects of the USAIA framework are described in detail in [183, 223].

The mobility management at the cell level is performed using a new fast handoff protocol. Furthermore, USAIA provides all necessary mechanisms for resource reservation in advance. For this purpose, the signaling protocol RSVP (Resource Reservation Protocol [22]) is modified in such a way that mobile nodes (MNs) are able to request resources on the BSs to which they have not yet performed a handoff. These requested resources are called passive reservations, because they can be used by the best-effort traffic of other MNs, as long as no handoff of the requesting MN has occurred. After a successful handoff, passive reservations turn into active ones. In that state, they cannot be used any longer by other traffic. To distinguish between these kinds of resources, the available bandwidth of a BS is partitioned, as depicted in Fig. 5.1.

Reservations of MNs requesting new resources from the network fall into the initial reservation class. Active and passive reservations of users leaving the scope of their initial or current BSs fall into the continuous reservation class. Reservations of MNs sharing the best-effort delivery traffic fall into the no reservation class. The mechanism to request resources in advance minimizes possible distortions of the real-time transmission during handoffs. This can be assured as long as the BSs are able to accept passive reservations. To provide the opportunity to accept as many passive reservations as possible,

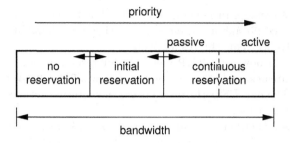

**Fig. 5.1.** Bandwidth partitioning

the protocols involved are optimized and extended in several ways. The details of these optimizations can be found in [183].

The key aspect of the USAIA framework is the seamless interworking of the mobility management and the resource reservation mechanisms to provide appropriate support of real-time traffic. To understand the interactions of both mechanisms, the message exchange between a correspondent host (CH) and the MN via the BS is explained in more detail (see also Fig. 5.2).

- The initial message is the beacon signal of a BS, containing the IP address of the BS that sends it. Beacon signals are sent via broadcast. The rate of the beacon signals is correlated to the current load and the available resources of the BS. To support fast handoffs, the beacon signal conveys a provider-defined share of the maximum resources that can be requested by MNs. This happens either during the handoff procedure as a passive reservation or as an initial reservation, i.e., after the successful termination of the authentication process, when the MN enters the network.

- Within the cell overlap areas, the MN recognizes beacons from different BSs, and triggers the handoff procedure. The handoff is initiated by sending an MN_Announce message to the new BS, carrying the MN's own address as well as that one of the "old" BS. The MN_Announce message contains either an indication of whether the handoff is marked as "triggered" or whether it is a "handoff announcement." The former initiates the handoff procedure and the latter makes it possible to request resources in advance, without losing connectivity to the current BS.

- The new BS confirms receipt of the MN_Announce message with an MN_Announce_Ack message.

- If the MN_Announce message is marked as triggered, the new BS sends a Notify message via the wired link to the old BS to inform it that the MN has moved. This message conveys the address of the new BS. The new BS also creates a routing table entry for the MN.

- Receiving a Notify message, the old BS deletes its routing table entry for the MN.

- If the MN_Announce message is not marked as triggered, the BS accepts a retransmitted MN_Announce message to trigger the handoff or the reservation in advance requests, respectively.
- The MN can now send RSVP messages for reservations in advance; these messages are acknowledged by the new BS by a RSVP RESV Confirmation message. The new BS performs all necessary internal actions to handle this passive reservation.
- The MN sends either a retransmitted RSVP RESV message to maintain its passive reservation(s) or a retransmitted MN_Announce message marked as triggered. The latter initializes the real handoff.
- On receiving a triggered MN_Announce message, the new BS turns the passive reservation into an active reservation and transmits a Notify message to the old BS.
- Finally, the new BS broadcasts a gratuitous proxy ARP (Address Resolution Protocol) to map the MN IP address to the BS link layer address, thus forcing all nodes involved to update their ARP caches with that information. This mechanism prevents the chaining of several BSs.

The most critical point of the USAIA framework with respect to real-time traffic support is the mechanisms provided on the cellular level due to the following reasons. First, the resources on the wireless link are usually much more limited than on the wired part of the network. Second, handoffs between

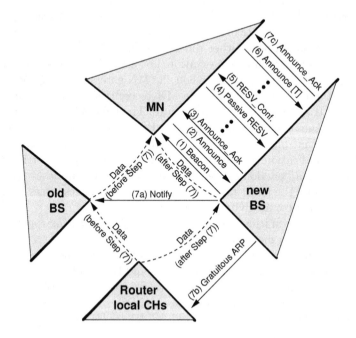

**Fig. 5.2.** Local handoff protocol

cells will likely happen more frequently than on the domain level or global level, thus influencing real-time traffic more significantly.

The major goal of the USAIA framework is the decoupling of the MN movement from the resources provided on any location visited. This means that there should be no need to stop the MN movement just to avoid any distortion of its real-time traffic. Therefore, the performance evaluation of all traffic classes involved with respect to accepted and failed reservations at the cellular level is of major interest.

## 5.2 Petri Net Model

Following the guidelines of Chap. 4, the modeling of the USAIA framework is started with the fastest and most simple model development scheme: the Petri net description. Another guideline of Chap. 4, that can be applied here, is to profit from symmetry. The Petri net description is based on modeling events of a single BS transmission range assuming a symmetric cellular network (as in Fig. 2.36) and that all BSs behave the same way. The model represents the transmission range of the BS and all MN activities that might occur within that range. These activities include nodes arriving at the range of the BS, nodes moving within the range, nodes initializing a new association within the range (MNs that are "switched on"), and nodes leaving the range. The term association denotes any kind of connection between BS and MN.

With regard to the kind of traffic MNs deal with, the model distinguishes MNs that carry real-time traffic (in the context of USAIA, this is related to a previous setup of a reservation request via RSVP) and MNs carrying best-effort traffic. Furthermore, the model takes into account MNs that establish a new association within the range of the BS. Any MN that deals with more than a single data flow within an association is handled by the model by dividing this node into multiple nodes, each one dealing with one of the data flows within a separate association. For instance, an MN carrying a data flow with real-time traffic and a data flow with best-effort traffic is represented by two MNs: one of them carrying a data flow with the real-time data and the other with the best-effort traffic. According to the guidelines of Chap. 4, a complex data flow scenario is decomposed into two simpler ones. A state space reduction results because only single data flows occur, and no combined data flows have to be represented by additional states.

To deal with the broad scope of individual reservation requests (initial and passive) of arbitrary size up to the permitted limit of bandwidth, discretization is applied as proposed in Chap. 4; the real-time traffic of MNs is assumed to belong to one of the three following classes: a class of low bandwidth requirements (called Min), a class of high bandwidth requirements (called Max), representing the maximum share of bandwidth a provider is willing to accept for a single request, and a class of average bandwidth requirements (called

`Avg`). Each of those classes is divided into two subclasses according to the extension of the Controlled Load Service Class described in [183].

Subclass `R` indicates that the requested bandwidth of a passive reservation can be reduced by the BS in the case of further incoming passive reservation requests and if the number of accumulated bandwidth of passive reservations exceeds a certain threshold value. For simplicity, our model distinguishes only between reservations that can be reduced to the lowest level of service the application is willing to accept and reservations that cannot be reduced due to the missing tolerance range. The latter means that the range specified in the Controlled Load Service specification of the application is set to 0. In reality, USAIA deals with the entire range of possible reductions, because the BS uses only the appropriate percentage of the tolerance ranges of all passive reservations to admit a new passive reservation request.

Subclass `F` means a fixed required bandwidth for passive reservations with no reduction allowed. As a result, different bandwidth allocations are modeled by six subclasses: `MinF`, `AvgF`, `MaxF`, `MinR`, `AvgR`, and `MaxR`.

In consequence, applying the discretization guideline of Chap. 4 to the bandwidth in the USAIA framework keeps the model tractable, while using continuous bandwidth would lead to an infinite state space. Of course, the more the number of bandwidth classes, the more accurate the model. In the following example, it turns out that three classes combined with two subclasses are sufficient to perform the required investigations.

DSPNs are used to describe the USAIA framework. Any passing time is represented by transitions that consume time from enabling to firing. Transitions with deterministic firing times are modeled by black filled rectangles. Transitions with exponentially distributed firing times are modeled by white rectangles. Immediate transitions, which do not consume any time, are modeled by black bars. All model input parameters, that can be changed are given in Table 5.1. They are explained in more detail in the following.

Each token appearing in the model represents an MN, except tokens in place `Beacon` modeling the beacon signal. Therefore, a kind of token "flow" through the net occurs, representing the current state of the corresponding MN. To handle multiple MNs (tokens) in parallel, all timed transitions are infinite server transitions except `T1` (see Fig. 5.4), which models the MN arrival at the BS, and `T44` (see Fig. 5.3), which models newly started applications at MNs.

The Petri net is described in detail below, starting with MNs opening their communication with an initial association. For simplicity, the authentication process is not taken into account.

### 5.2.1 Initialized Mobile Nodes

MNs that stay in the range of the BS and that like to establish a new association with a BS due to a newly started application are modeled by transition

**Table 5.1.** Model parameters

| Parameter | Meaning |
|---|---|
| DnewMN | average time between new node arrivals |
| Dinit | average time between new initializations |
| Doverlap | time to pass the overlap area |
| Dremain | time to pass the remaining range |
| Dprocess | time for handshake between BS and MN |
| Dbeacon/DbeacInc | base/decrement time of beacon |
| fracRTT | fraction of real-time traffic |
| AvgRcvdBS | average number of received BSs |
| frac*Class* | traffic fraction of class *Class* (Table 5.2) |
| guard T49 etc | accepting an active reservation |
| guard T40 etc | threshold of passive bandwidth reduction |
| guard T45/T46 etc | accepting a passive reservation |
| guard T4 etc | accepting a initial reservation |

T44 (Fig. 5.3). The rate of new applications requesting an association is assumed to be exponentially distributed. The average time between two requests is identified by the parameter Dinit. A new association to be established (token in place Init) requests either best-effort traffic or real-time traffic. The relation between the two kinds of traffic is modeled by the weights of the transitions T42 and T43. The parameter fracRTT defines the ratio of real-time traffic, and, therefore, defines the weight of transition T43. The weight of transition T42 results in 1-fracRTT.

If the MN requires real-time traffic (firing of transition T43), the node listens for a beacon signal (place listInitRT). The beacon signal model is introduced in Sect. 5.2.2. If such a signal is received (firing of transition T39), the node announces its presence to the BS (place AnnInit) by an MN_Announce message. Transition T38 models all the time needed for that procedure. That is the announcement procedure itself, including local processing time, and the reservation request procedure, also including the local processing time. Stochastic influence on this time, for instance, additional delays due to signal interferences, is neglected to keep the model simple. Thus, this time is deterministic and expressed by parameter Dprocess. This parameter is assumed to be independent of the kind of reservation (initial or passive).

The model provides dealing with network architectures where multiple BSs are in range of the MN. The parameter AvgRcvdBS defines the average number of BSs that are received at an MN's location. Of course, an initial association will be set up to only one of those BSs. Transition T36 ensures that the BS of our model will get the initial association. Otherwise, the request is turned to a passive reservation (firing of transition T37). The weight of T37 results in (AvgRcvdBS - 1)/AvgRcvdBS. Such an initial passive reservation is treated similarly to passive reservations that result from such MNs arriving at the

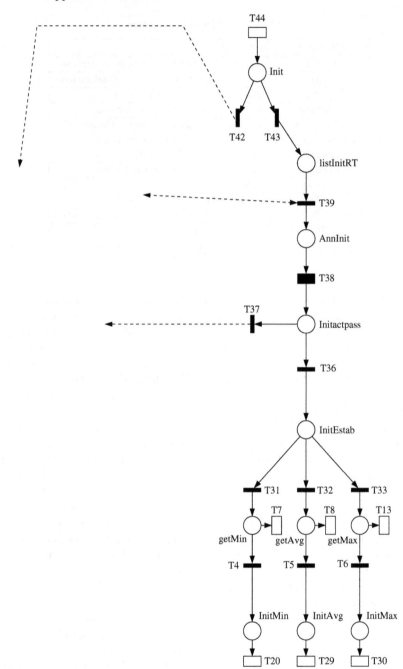

**Fig. 5.3.** MNs establishing a new association

range of the BS (Sect. 5.2.2). This means such passive reservations may turn into active ones (e.g., when the initial reservation to the other BS terminates), considering the same assumptions (concerning path, and so on) as in the case of arriving nodes.

The firing of transition T36 indicates that the initial reservation is set up at the modeled BS. The weight of T36 results in 1/AvgRcvdBS. Because initial reservations are also assumed to belong to one of the three bandwidth classes Min, Avg, and Max, the transitions T31, T32, and T33 manage the distribution among the classes by their weights. The parameter fracMinI (weight of transition T31) gives the fraction of traffic class Min, fracAvgI (weight of T32), the fraction of class Avg, and fracMaxI (weight of T33), the fraction of class Max. The MN will try to establish a reservation until the required bandwidth is accepted or the MN leaves the range of the BS.

If the traffic belongs to the class Min, a token is generated in place getMin. Establishing the corresponding reservation is done by transition T4. But T4 is only enabled if its guard is fulfilled:

$$\#\text{InitMin} + \#\text{InitAvg} \cdot \text{AvgMult} + \#\text{InitMax} \cdot \text{MaxMult}$$
$$\leq \text{Initbw} - 1. \tag{5.1}$$

The guard prevents a new reservation if the required bandwidth is not available. AvgMult defines the factor of the bandwidth the class Avg requires relative to the class Min. MaxMult defines the factor of the bandwidth the class Max requires relative to the class Min. Initbw represents the total bandwidth of the BS for initial reservations relative to the bandwidth of the class Min.

If the guard is fulfilled, an initial reservation is established (token in InitMin). The time the MN stays within the range of the BS is modeled by the exponentially distributed transition T20 with an average delay of (Doverlap + Dremain)/2. Given such a delay, it is assumed that the MNs are initialized while spending half of their average time in the range of the BS. Doverlap and Dremain will be defined in Sect. 5.2.2. Due to the lack of movement patterns of MNs, the firing delay and its distribution are assumed. Besides exponential distribution, any other kind of distribution could also be modeled using a transition with a generalized distribution.

If the guard is not fulfilled, the MN tries to establish an initial reservation as long as it is in the range of the BS. Leaving the range is modeled by transition T7 with an average delay time of (Doverlap + Dremain)/2. The measure RInitMin = E(#InitMin) gives the average number of initial reservations, and RIgMin = E(#getMin) gives the average number of nodes waiting for an initial reservation.

Both of the other bandwidth classes are modeled in a similar way. The only difference is given by the bandwidth requirements, which result in slightly different guards concerning the corresponding transitions to T4. For instance, in the case of the bandwidth class Avg, the maximum occupied bandwidth at the BS (right-hand side of Eq. (5.1)) has to be changed to Initbw - AvgMult because the newly established association requires `AvgMult` times the minimal bandwidth (of class `Min`).

### 5.2.2 Real-time Traffic

MNs that enter the transmission range of the BS are modeled by the transition `T1` (Fig. 5.4). The arrival rate is assumed to be exponentially distributed. The average time between two arrivals is given by the parameter `DnewMN`, which can be set depending on the MN density and movement pattern. The relation

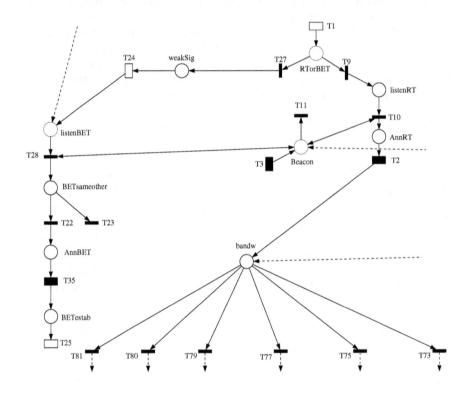

**Fig. 5.4.** MNs entering the BS range

between real-time traffic and best-effort traffic is modeled by the weights of the transitions `T27` and `T9`, as in case of initialized MNs.

An arriving MN first enters the overlap area of the new and the old BS. If it carries real-time traffic (firing of transition `T9` in Fig. 5.4), the node listens for a beacon signal (place `listenRT`) because it is willing to establish an association for a passive reservation. Transition `T3` represents the generation of the beacon signal, and therefore fires at a predefined fixed rate. Firing leads to a token in `Beacon`. The MN receives the signal (firing of transition `T10`). Other kinds of traffic are also served in this way (transitions `T28`, best effort traffic, and `T39`, initialized nodes; Fig. 5.3). These three transitions are preferred to transition `T11` due to their higher priority, set to 2. A firing of one of the three transitions does not lead to a deletion of the token in place `Beacon` because the corresponding arcs are double-sided. As a result, all MNs listening are handled first, and then the token representing the beacon signal is removed by the low priority transition `T11`.

The rate of `T3`, i.e., the rate of the beacon signal, is usually load dependent. This means that the higher the BS load (bandwidth occupied by associations), the lower the rate. As a result, the delay time of `T3` is a marking-dependent function that deals with this constraint. Here, the delay time is controlled by the number of associations. It results in a basic rate of 1/`Dbeacon` in the case of no association. `Dbeacon` represents a model parameter. The rate is decreased for each association depending on a second parameter, `DbeacInc`. Other marking-dependent functions can be chosen as well.

If a beacon signal is received (firing of transition `T10`), the node announces its presence to the BS (place `AnnRT`) and requests a passive reservation via RSVP. Transition `T2` accumulates the time required to handle this request, including the time for message passing, processing, and so on, similarly to `T38` in the case of initialized MNs. The delay of `T2` is given by the parameter `Dprocess`. The remaining model only considers the MN movement time.

After handling the request for a passive reservation, a token in place `bandw` occurs. Six transitions are enabled now (`T81`, `T80`, `T79`, `T77`, `T75`, and `T73`). They represent the assignment of the traffic carried by the MN to the six different real-time traffic bandwidth classes. The ratio of class `MinF` is modeled by the weight of transition `T81` and determined by parameter `fracMinF`. Table

**Table 5.2.** Bandwidth classes

| Class | Transition | Weight parameter |
|-------|------------|------------------|
| MinF  | T81        | fracMinF         |
| AvgF  | T80        | fracAvgF         |
| MaxF  | T79        | fracMaxF         |
| MinR  | T77        | fracMinR         |
| AvgR  | T75        | fracAvgR         |
| MaxR  | T73        | fracMaxR         |

5.2 gives the transitions and weights of all bandwidth classes. After the firing of one of the transitions, a token is generated in the corresponding subnet, as shown in Fig. 5.5. The token in place `bandw` is removed. The token "flow" within the subnet is exemplary explained for the bandwidth class `MinF` of Fig. 5.5.

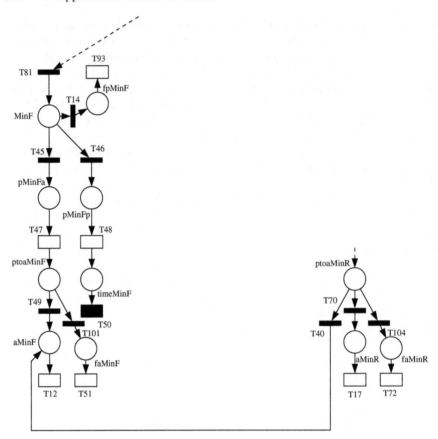

**Fig. 5.5.** One of the bandwidth classes

A token in place MinF indicates that the MN is carrying traffic of the class MinF. Next, the load of the BS is investigated and it is determined whether a passive reservation to the MN can be established (firing of transition T45 or T46) or whether the request for a passive reservation is to be denied due to the lack of bandwidth (firing of transition T14).

Compared to the priority 1 of transition T14, the transitions T45 and T46 have a higher priority (priority of 2), and therefore, establishing the passive reservation has priority over denying it. But transitions T45 and T46 are also guarded. Both guards are equal and guarantee that a passive reservation is only established if the available bandwidth allows turning it into an active reservation if necessary. Nevertheless, the model also allows reserving more bandwidth than is available because not all passive reservations are turned into active ones. Therefore, the previously mentioned guards are of free choice, and the Petri net model may help find the optimal guards and the optimal constraints in the real network for accepting passive reservations. Examples of guards are found in Sect. 5.3.

If T45 and T46 are not enabled due to the guards, T14 is fired and places a token in place fpMinF: the passive reservation has failed. The MN crosses the range of the BS without any reservation to it. The crossing time is modeled by transition T93 assuming an exponential distribution (or any other kind of distribution modeled by transitions with generalized distributions) with an average crossing time of Doverlap + Dremain. Doverlap models the time an MN needs to cross the overlap area of the old and the new BS. The time to traverse the remaining (new) range is given by Dremain. The average number of tokens in fpMinF determines the average number of failed passive reservations RfpMinF = E(#fpMinF).

If T45 and T46 are enabled, a passive reservation can be established. As mentioned in Sect. 5.2.1, the model allows dealing with network architectures where multiple BSs are in range of the MN. The parameter AvgRcvdBS defines the average number of BSs that are received at an MN's location. Of course, an active association will later be set up to only one of those BSs. Transition T45 ensures that the BS of our model will get the active association, and transition T46 ensures that one of the AvgRcvdBS − 1 other BSs will get it. Due to the assumed uniform movement pattern, the weight of transition T45 results in 1/AvgRcvdBS, and the weight of transition T46 results in (AvgRcvdBS − 1)/AvgRcvdBS.

If T46 fires, the MN will never establish an active reservation at the modeled BS. Place pMinFp represents the MN crossing the range of the BS. Transition T48 determines the average crossing time, given by Doverlap + Dremain. Instead of the exponential distribution, any other one may be chosen. Place timeMinF represents that the MN has left the transmission range. The passive reservation times out after the elapsed time defined by transition T50. It is given by the parameter Dtimeout. The average number of tokens in timeMinF allows determining the number of passive reservations that are timing out: RToutMinF = E(#timeMinF).

If T45 fires, the MN will establish an active reservation. Place pMinFa models the MN crossing half of the overlap area of the old and the new BS (modeled by an average delay time of Doverlap/2 by transition T47). Usually, the signal of the old BS will become weaker than the signal of the new BS after the crossing of half of the overlap area. As a result, the firing of T47 and a token in place ptoaMinF means that the passive reservation has to be turned into an active one: transition T49 fires if the available bandwidth of the BS is sufficient (a token is placed in aMinF). Otherwise, the firing of transition T101 indicates the failure of turning to an active reservation (token in place faMinF). The guard of T49 and the priorities of both transitions deal with this decision. Transition T49 has the higher priority, but also a guard in the way of

$$\#\texttt{aMinF} + \#\texttt{aAvgF} \cdot \texttt{AvgMult} + \#\texttt{aMaxF} \cdot \texttt{MaxMult}$$
$$+ (\#\texttt{aMinR} + \#\texttt{aAvgR} \cdot \texttt{AvgMult} + \#\texttt{aMaxR} \cdot \texttt{MaxMult})/\texttt{reduce}$$
$$\leq \texttt{Contbw} - 1, \tag{5.2}$$

where `AvgMult` defines the factor of the bandwidth that the class `Avg` requires relative to the class `Min` (for both subclasses, `F` and `R`). `MaxMult` defines the factor of the bandwidth that the class `Max` requires relative to the class `Min`. `Contbw` represents the total continuous reservation bandwidth of the BS relative to the bandwidth of an association of class `MinF`. The parameter `reduce` models the factor of maximum bandwidth reduction of the reducible bandwidth classes. The guard in Eq. (5.2) ensures that the bandwidth for a new active reservation of the class `MinF` is available.

If the guard is fulfilled, the active reservation is activated (token in `aMinF`). The MN crosses the second part of the overlap area and the remaining range of the BS (modeled by the exponentially or alternatively distributed transition `T12` with an average delay of `Doverlap/2 + Dremain`). If the guard is not fulfilled, an active reservation fails (modeled by place `faMinF` and transition `T51` with an average delay of `Doverlap/2 + Dremain`). The measure `RaMinF` $= E(\#\texttt{aMinF})$ denotes the average number of active reservations and the measure `RfaMinF` $= E(\#\texttt{faMinF})$ determines the average number of failed active reservations.

All bandwidth classes are modeled in a way similar to that in Fig. 5.5. The only difference is given by the bandwidth requirements, and results in slightly different guards concerning the corresponding transitions to `T45`, `T46`, and `T49`. Concerning transition `T56` in the case of the bandwidth class `AvgF`, for instance, the maximum occupied bandwidth of the BS (right-hand side of Eq. (5.2)) should be `Contbw−AvgMult`.

The bandwidth classes `MinR`, `AvgR`, and `MaxR`, which represent passive reservations of reducible bandwidth, deal with an additional transition, as compared to the previously described case of fixed bandwidth (e.g., `T40` of bandwidth class `MinR` in Fig. 5.5). It models establishing an active reservation of non-reduced bandwidth if the BS load is light. The priority of `T40` is higher than those ones of `T70` and `T104`. But its guard prevents its firing if the BS is heavily loaded. The guard belongs to the model parameter set.

### 5.2.3 Entire Model

Combining the submodels of Figs. 5.3 to 5.5 lead to the entire Petri net model of the USAIA framework. It is depicted in Fig. 5.6. Table 5.3 shows all transition weights that differ from 1. The model development and evaluation was performed using the toolkit *TimeNET* [65].

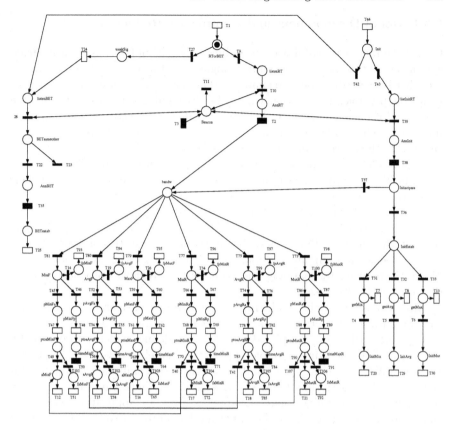

**Fig. 5.6.** Petri net model of the USAIA framework

**Table 5.3.** Transition weights

| Transition | Weight |
|---|---|
| T9,T43 | fracRTT |
| T27,T42 | 1 - fracRTT |
| T22,T36,T45 etc | 1/AvgRcvdBS |
| T23,T37,T46 etc | (AvgRcvdBS - 1)/AvgRcvdBS |
| T81 etc, T31 etc | fracClass |

## 5.3 Model Engineering and Performance

The model previously presented is revisited here in order to give a summary of how the design rules from Chap. 4 have influenced the architecture.

### 5.3.1 Model Development and Complexity Reduction

One of the most successful ways to reduce model complexity is to profit from symmetry, as this model does. Only a single base station is modeled, assuming all other base stations behave similarly. The data representation chosen is as simple as possible. Tokens that cannot be distinguished represent the mobile nodes and their movements. Other MN information, such as an identification number or billing information, is not needed to determine the performance of the USAIA framework for the given questions.

The model also combines events. Not all steps of the handoff protocol as given in Fig. 5.2 are explicitly incorporated. For instance, requests and their acknowledgments are combined in a single transition that reflects the amount of time needed for communication.

Discretization is applied by introducing only three bandwidth classes. The bandwidth that the MNs request in reality will usually be continuously distributed between minimum and maximum bandwidths. Here, the model complexity is reduced by mapping the requested bandwidth onto one of three discrete bandwidth classes.

However, some of the guidelines have not been considered. For instance, the computational overhead of a Petri net description is taken into account to come up with a fast graphical high-level modeling, which is provided by Petri nets. The following example applies the model to a wireless local area network (WLAN) environment and shows that the computation time is still small enough to perform all desired investigations. Thus, any other modeling techniques can be waived here. Chapter 6 gives an application where the simulation performance of the Petri net model is prohibitive for the required analysis of the system. In this case, other modeling techniques have to be used.

### 5.3.2 Modeling Power

The previously presented model is independent of cellular network technology. For each specific technology, the parameters can be adapted.

To demonstrate this model, the parameter set of a wireless local area network (WLAN) [27, 31, 64, 174, 236] scenario is used, according to IEEE 802.11b.

The features of a commercial product are applied as an access point, which in this case is the realization of the BS. Both terms are synonymous. For the model, it does not matter that access points are not IP-aware, because all time constraints were taken into account as if they were processing IP packets.

The diameter of the transmission range is about 60 meters for data rates up to 11 Mbps. Users (MNs) are assumed to pass the transmission range on paths that cover a length of half the diameter on average. If they walk (7 km/h), they spend 15.429 seconds within the range. It is further assumed that the network is set up in such a way that an average overlap of 10 meters occurs along the path of a user. Of course, the model can also deal with any other overlap. As a result, the overlap area of 10 meters is passed in `Doverlap` = 5.143 s and the remaining path within the transmission range in `Dremain` = 15.429 s − 5.143 s = 10.286 s.

The BS is assumed to process any announcements or reservations of MNs in `Dprocess` = 0.001 s. Passive reservations time out after `Dtimeout` = 15 s. The beacon signal is initially sent at a rate of 10 per second (`Dbeacon` = 0.1 s) and decreased for each 10 established connections (`DbeacInc` = 10) by one.

MNs are assumed to be located in the range of `AvgRcvdBS` = 2 BSs on average. Their movement pattern and density result in MNs that arrive at a new BS at a rate of 1 per `DnewMN` = 0.05 s. MNs are initialized within the range of a BS at a rate of 1 per `Dinit` = 0.15 s. Half of all new nodes are assumed to carry best-effort traffic, and half of them real-time traffic (`fracRTT` = 0.5).

According to the product specification, the base station is able to handle about 50 nominal users that require a small bandwidth. Those users are associated with our bandwidth class `MinF` (and `Min` initial reservations). The mainstream users are assumed to consume twice the bandwidth of nominal users, and are associated with bandwidth classes `AvgF` and `Avg`. Power users are assumed to require four times the bandwidth of nominal users, and belong to the classes `MaxF` and `Max`. Users that can reduce their consumed bandwidth are assumed to allow a maximum reduction of 50%. The BS bandwidth is divided into a bandwidth of a maximum of 40 nominal users for active reservations and into a bandwidth of a maximum of eight nominal users for initial reservations. All bandwidth classes are assumed to occur with equal probabilities.

Concerning the guards that observe whether passive reservations can be turned into active ones (e.g., Eq. (5.2)), previous definitions lead to `AvgMult` = 2 and `MaxMult` = 4. A maximum reduction of 50% results in `reduce` = 2. Thus, the guards of transitions T49, T56, T63, T70, T83, and T90 are determined. Furthermore, the guards that observe whether initial reservations are accepted are also set up by their corresponding equations (e.g., Eq. (5.1)).

The guards that observe whether a passive connection can be established are determined in a similar way. Nevertheless, any other arbitrary enabling function can also be chosen. The guard we have chosen considers all active reservations and their corresponding bandwidths. Only a fraction of the mentioned active bandwidth reservations is taken into account. This is because usually some time passes until a passive reservation turns into an active one. During this time, some of the other active reservations of the BS are terminated because the corresponding MNs have left the transmission range of the BS or terminated their applications. Thus, only the remaining fraction of active reservations is considered by the guard. The parameter `remact` allows assuming this fraction. The example presented sets it to 60% (`remact` = 0.6).

Furthermore, the passive reservations and their corresponding bandwidths (if they turn into active reservations) are considered. Only a fraction of those reservations is taken into account because several passive reservations will time out and never turn into an active reservation. This ratio is given by the parameter `rempass`. It is the investigated parameter of the given example: What ratio should be chosen to achieve a high number of active and passive reservations but also a low number of failed ones? The guards of transition T45 and T46 result in

$$
\begin{aligned}
0.6(&\#\text{aMinF} + \#\text{aAvgF} \cdot 2 + \#\text{aMaxF} \cdot 4 \\
&+\#\text{aMinR} + \#\text{aAvgR} \cdot 2 + \#\text{aMaxR} \cdot 4) \\
+\text{rempass} \cdot ((&\#\text{pMinFp} + \#\text{timeMinF} + \#\text{pMinFa}) \\
&+(\#\text{pAvgFp} + \#\text{timeAvgF} + \#\text{pAvgFa}) \cdot 2 \\
&+(\#\text{pMaxFp} + \#\text{timeMaxF} + \#\text{pMaxFa}) \cdot 4 \\
&+(\#\text{pMinRp} + \#\text{timeMinR} + \#\text{pMinRa}) \\
&+(\#\text{pAvgRp} + \#\text{timeAvgR} + \#\text{pAvgRa}) \cdot 2 \\
&+(\#\text{pMaxRp} + \#\text{timeMaxR} + \#\text{pMaxRa}) \cdot 4) \leq 39. \quad (5.3)
\end{aligned}
$$

The guards of the other bandwidth classes differ only in the right-hand side of the equation that determines the maximum occupied bandwidth of the BS to allow at least one active reservation of the corresponding bandwidth class.

The BS is configured to deal with as many active reservations as possible. This means that traffic of subclass R is accepted by the BS only with their reduced bandwidth. It is modeled by disabling transitions T40, T41, and T107.

Figures 5.7 to 5.10 show some results. The reciprocal ratio of passive reservations is 1/`rempass`. The average number of active reservations (Fig. 5.7), the average number of failed active reservations (Fig. 5.8), the average number of passive reservations (Fig. 5.9), and the average number of failed active reservations (Fig. 5.10) are compared for all six bandwidth classes. The results are achieved by simulation due to multiple enabled deterministic transitions that cannot be mapped onto a Markov chain. As termination criteria, a confidence level of 95% and an estimated precision of 2% is used. But in the case of rare events, these criteria are relaxed due to a simulation run-time of more than 10 hours on a 1,200 MHz processor. Relaxing the estimated precision to 5% leads to about two hours simulation time.

The plots of the bandwidth classes MinF and AvgR, and those of AvgF and MaxR, are equal due to equal probabilities of the bandwidth classes and equal bandwidth of both classes if bandwidth reduction is taken into account.

The minimal ratio of passive connections to avoid failed active reservations arises as another result of the example. The reciprocal ratio should not be larger than 10 (Fig. 5.8), resulting in `rempass` = 0.1.

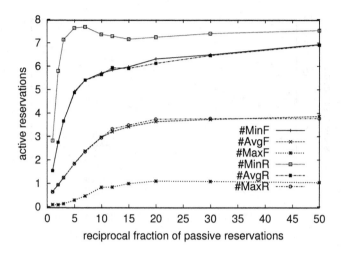

**Fig. 5.7.**  Active reservations

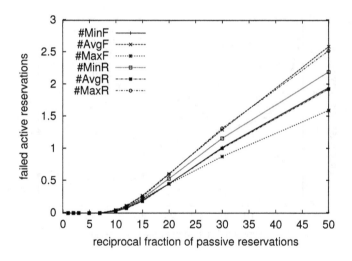

**Fig. 5.8.**  Failed active reservations

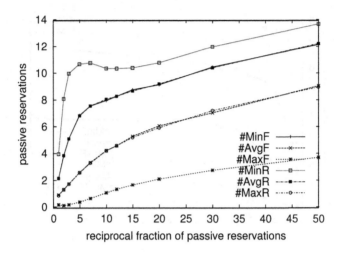

**Fig. 5.9.** Passive reservations

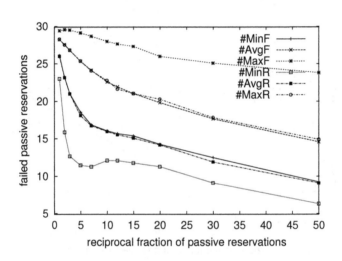

**Fig. 5.10.** Failed passive reservations

# 6

# Application: Multistage Interconnection Network

This chapter will give a second example of model engineering. Multistage interconnection networks serve as the network architecture in question. This example is much more exhaustive than the previous one. Simulation models as well as mathematical models are considered.

Multistage interconnection networks (Sect. 2.3.10) are often proposed to connect a parallel or distributed system. MINs with the banyan property performing packet switching will be of special interest. Again, as in the previous example of modeling bandwidth partitioning for cellular networks, Petri nets are developed first as they allow easy and fast model construction.

Petri nets allow solving the model by mathematical methods (analyzing the underlying Markov chain) or by simulation. But in the example presented, it will turn out in Sect. 6.1 that mathematical solutions using Petri nets are not feasible, and only simulation can be applied. Besides Petri net simulation, a particularly adapted C++ simulation as well as Markov chains will be used to model the MIN.

## 6.1 Simulation: Petri Nets

To gain experience in modeling MINs, some assumptions are made to keep the model as simple as possible:

- The traffic load (offered load) to all inputs of the network is equal. It is assumed that a packet is offered to each network input at each clock cycle.
- Packet destinations are uniformly distributed. This means that each output of the network is with equal probability one of the destinations of a packet.
- Conflicts between packets for network resources are randomly solved with equal probabilities.
- The switching of all elements is synchronously performed with an internal clock cycle.

- Routing is performed in pipeline manner. This means that the routing process occurs at every stage in parallel.
- All packets have the same size (like in ATM).
- Packets are immediately removed from their final destinations after arrival.
- The destinations of succeeding packets are independent of each other.

These assumptions allow establishing a model to determine the performance of $N \times N$ MINs consisting of $2 \times 2$ SEs with $n = \log_2 N$ stages. At each stage $k$ $(0 \le k \le n-1)$, there is a FIFO buffer to store a single packet in front of each SE input. The packets are routed by store-and-forward switching from one stage to its succeeding one by the backpressure mechanism.

### 6.1.1 Full Petri Net Model

Figure 6.1 shows the Petri net description of an $8 \times 8$ MIN consisting of $2 \times 2$ SEs [212]. The Petri net behavior is determined by the behavior of twelve

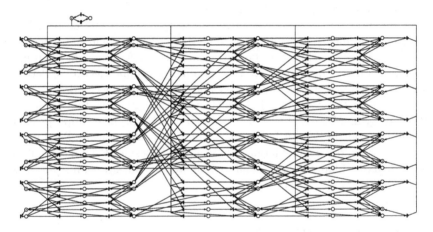

**Fig. 6.1.** Petri net model of an $8 \times 8$ MIN

$2 \times 2$ switching elements. One of them is shown in Fig. 6.2. Table 6.1 gives the priorities and weights of the immediate transitions. The only timed transition is called CycleTime. It represents (together with places NewCycle and FinishCycle and with transition ResetCycle) the clock cycle of the network. It is, of course, a deterministic value.

   Places Buf1 and Buf2 model the two input buffers of the switching element. Places Buf1empty and Buf2empty show the states of the two buffers. If there is a token in a place, the corresponding buffer is empty. The packet is represented by a token in Buf1 or Buf2, respectively. If a packet enters the switch at the upper input port, a token is placed in place Buf1, and the token in place Buf1empty is removed via the Petri net description of the SE of the preceding

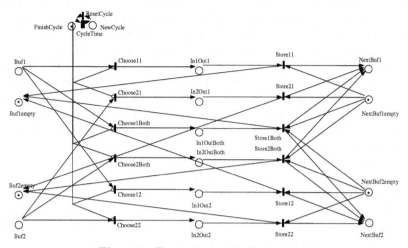

**Fig. 6.2.** Petri net model of a 2×2 SE

**Table 6.1.** Transitions of an SE

| Immediate trans. | Priority | Weight | Description |
|---|---|---|---|
| ResetCycle | 2 | 1.000000 | clock |
| Choose11 | 4 | $0.5 \cdot \omega_1$ | |
| Choose21 | 4 | $0.5 \cdot \omega_1$ | selection |
| Choose1Both | 4 | $\omega_2$ | of the |
| Choose2Both | 4 | $\omega_2$ | required |
| Choose12 | 4 | $0.5 \cdot \omega_1$ | output |
| Choose22 | 4 | $0.5 \cdot \omega_1$ | |
| Store11 | 1 | 1.000000 | transmission |
| Store21 | 1 | 1.000000 | to the |
| Store1Both | 1 | 1.000000 | input |
| Store2Both | 1 | 1.000000 | buffer |
| Store12 | 1 | 1.000000 | of the |
| Store22 | 1 | 1.000000 | next stage |

stage or via the Petri net description of the preceding node if it is the first
stage. The transitions Choose11, Choose1Both, and Choose12 are inhibited
because there is a token in place FinishCycle. When the clock cycle starts,
transition CycleTime fires after the deterministic cycle time. In the new state,
one out of the three transitions Choose11, Choose1Both, or Choose12 fires
because they have priority 4, in contrast to transition ResetCycle, which has
priority 2. Transition Choose11 will fire if the upper output of the SE is the
destination of the packet. Transition Choose12 will fire if the lower output of
the SE is the destination, and transition Choose1Both will fire if both outputs
are the destination. The weights $\omega_1$ and $\omega_2$ of these transitions determine with

what probability only a single output (which can be the upper or the lower one) or both outputs are the destination of the packet, respectively. They can be chosen in relation to the desired packet multicast traffic pattern [207]. The calculation of $\omega_i(k)$ giving the probability that $i$ SE outputs are the destination of a packet at the SE input is described for a given multicast traffic in Sect. 6.1.3. After firing, the token in place Buf1 is removed and arrives in place In1Out1, In1Out2, or In1OutBoth.

In the next step, transition ResetCycle fires because of its priority over all Store transitions. In the last step, the packet, represented by the token in place In1Out1, In1Out2, or In1OutBoth, will be moved to the input buffers of the corresponding SEs at the next network stage. These buffers are represented by places NextBuf1 and NextBuf2, and their states are represented by places NextBuf1empty and NextBuf2empty.

If there is a token in place In1Out1, the transition Store11 will fire depending on the availability of the upper output. It is available if the corresponding input buffer of the next stage is empty, represented by a token in place NextBuf1empty. A conflict with a packet from the lower input of the current switch, represented by a token in place In2Out1 or In2OutBoth, is randomly solved with equal probabilities because both enabled transitions, Store11 and Store21 or Store11 and Store2Both, respectively, have the same priority. If transition Store11 fires, the token is moved from place In1Out1 to place NextBuf1. The token in place NextBuf1empty is removed and place Buf1empty receives a token. This completes a clock cycle.

The case of a token in place In1Out2 is similar to the case above except that the lower output is the destination. If there is a token in place In1OutBoth, the transition Store1Both will fire if both outputs are available, represented by a token in place NextBuf1empty and place NextBuf2empty. This is called a complete multicast. Alternatively, a partial multicast may be modeled [212]. In such a scenario, a subset of the packet's copies are also allowed to proceed if not all desired outputs are available. In each case, a conflict with a packet from the lower input of the current SE is also randomly solved with equal probabilities.

The process just described occurs at each SE in the network at the same time and is repeated at each clock cycle: packets are moved from the input ports of the network through the stages to the output ports of the network.

Given the model of Fig. 6.1, performance results can be obtained. For instance, the normalized throughput of the network at the output (called $S_o$) can be determined by calculating the probability of a token being in the place modeling the related network output (rightmost places of Fig. 6.1).

Because tokens in DSPNs cannot be distinguished, additional places and transitions must be established to measure the delay time of packets in the MIN. Fig. 6.3 and Table 6.2 show how to model delay times. Measurement is performed in such a way that a new token is added to the model part for delay time measurement at each clock cycle (place MeasSt). If there is no token (i.e., no packet) in the measured input-output path of the SE, the token in MeasSt

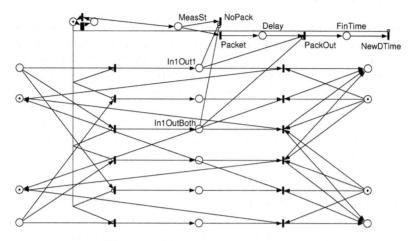

**Fig. 6.3.** Measurement of delay times

**Table 6.2.** Parameter of transitions (delay time measurement)

| Timed transition | Priority | Weight |
|---|---|---|
| NoPack | 3 | 1.000000 |
| Packet | 2 | 1.000000 |
| PackOut | 1 | 1.000000 |
| NewDTime | 4 | 1.000000 |

will immediately be removed by transition NoPack. On the other hand, if there is a token in the input-output path, the token is moved via transition Packet to place Delay. This place collects as many of such tokens as the packet spends clock cycles at this stage. All tokens are moved via PackOut to place FinTime when the packet leaves the stage. The number of tokens is then measured. At the end of the current clock cycle, they are removed by transition NewDTime.

Unfortunately, this Petri net model suffers from the huge state space it generates. Even for small networks, the state space is too large to be accommodated in computer memory. Therefore, a mathematical solution of the model is not feasible. Simulation can still be applied. However, simulation suffers from the drawback of high simulation run times. For instance, 64×64 MINs require more than two weeks of simulation if termination criteria are set to a confidence level of 95% and an estimated precision of 2% on a 1,200 MHz processor. Larger networks cannot be handled by this model.

### 6.1.2 Iterative Petri Net Model

To reduce model complexity, decomposition and symmetry considerations are applied. Profiting from decomposition and symmetry in the case of Petri net models of MINs requires measure-dependent transitions [211]. As fixed point iteration must be used to deal with cyclic dependences between the subsystems, one way to consider the results of the steady-state probability vector of the previous iteration is to adapt the parameters of the transitions in the Petri net. These transitions are said to be measure dependent. Two ways exist to introduce measure-dependent transitions: adapting the firing rate of a transition with exponentially distributed firing time and adapting the weight of an immediate transition. Both ways will be presented in the following.

In this derivation, it is assumed that the behavior of a timed Petri net can be divided into time intervals of similar actions (e.g., the Petri net description contains a clock cycle, as in the case of multistage interconnection networks). A time interval represents an iteration step. Thus, a transition and its probability $p_{fire}$ that it will fire if enabled can be used as the synchronization point between the subsystem models or as the synchronization point within a single model of a subsystem in the case of symmetry. During such a time interval, the probability $p_{fire}$ will depend on the steady-state probability vector of the previous iteration: $p_{fire}(i) = f_p(\pi(i-1))$, where $i$ is the iteration number. In the case of convergence, we get $p_{fire} = f_p(\pi)$.

To realize the firing probability within the time interval dependent on a subsystem measure, a transition of exponentially distributed firing time with a firing rate of

$$\lambda = -\frac{\ln(1 - p_{fire})}{T}, \tag{6.1}$$

resulting from $p_{fire} = 1 - e^{-\lambda T}$, is applied, where $T$ is the duration of the time interval. Such a transition can be chosen either to push new tokens into the model of the subsystem in question with the desired probability or to remove them from the model. The probability of pushing or removing a token is determined for each iteration step by considering the results of the other subsystem model and the dependences on it.

If the duration of the time interval $T$ is unknown or changes, for example, because it depends on the firing of a transition with exponentially distributed firing time, a synchronization of the subsystem models with another transition of exponentially distributed firing time is not possible. But the problem can be solved by using immediate transitions. Figure 6.4 shows the case of pushing a new token into the model. All transitions are immediate ones and have guards (square brackets). The Petri net description that models the subsystem to be examined starts with place ModelInp and is represented by dashed arcs. When the time interval starts, a token is in place Ready. If the model of the subsystem holds a state to accept a new token (a new token is permitted by the model via the guards), the transitions NewTk and NoNewTk are enabled. The weight of transition NewTk is $p_{fire}$, and the weight of transition NoNewTk

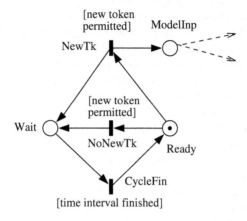

**Fig. 6.4.** Synchronization with immediate transitions (new token)

is $1 - p_{fire}$. Then, transition NewTk fires with probability $p_{fire}$, and with this probability (which depends on the other subsystem, or this subsystem itself in the case of symmetry), a new token enters the examined model, and another one is generated in place Wait. When firing transition NoNewTk, the token is only moved to place Wait. It stays there for the rest of the time interval. Now, transition CycleFin is able to fire, and the token moves back to place Ready. A new time interval starts.

Figure 6.5 shows the case of removing a token. The place ModelOutp represents the output place of the subsystem model from where the tokens should be removed with probability $p_{fire}$. This probability is the weight of transition

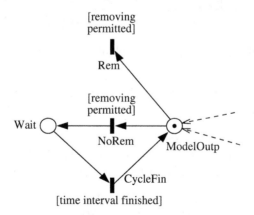

**Fig. 6.5.** Synchronization with immediate transitions (remove token)

Rem. Transition NoRem is weighted by $1 - p_{fire}$. The behavior within a time interval is almost identical to the previous case.

The probability $p_{fire}$ is updated at the end of each iteration step. An iteration step consists of the analytical solution of all subsystems or, if such a model has too many states to be solved analytically, it consists of the solution by simulation. In the case of simulation, the complete iteration process can be divided into precision steps. At each precision step, a new iteration run is started. The initial values are the corresponding results from the previous iteration run. The iteration run stops if the termination conditions are reached. For the early precision steps, the termination conditions of the simulation are chosen relaxed in order to accelerate the simulation of an iteration step. The closer the iteration converges to the fixed point, the more the termination conditions of the simulation are tightened to reach higher result precisions. The number of precision steps can be adapted to the model examined. However, at each iteration step, a simulation run must be performed. Even with relaxed termination criteria, this will usually consume more time than a single simulation run if the entire system is modeled and fixed point iteration is avoided. Therefore, the main focus of decomposition and symmetry is mathematical modeling methods.

Figure 6.6 serves as an example of decomposition and symmetry in the case of Petri nets. It shows a subsystem model of a MIN, which has already been modeled as an entire MIN, as depicted by Fig. 6.1. Only a single row of 2×2 SEs is modeled. The dependences on the other rows are realized by measure-dependent transitions with exponentially distributed firing times. Such tran-

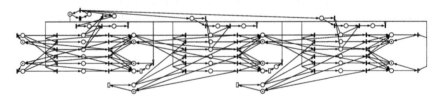

**Fig. 6.6.** Petri net description of an 8×8 MIN using decomposition

sitions implement the packet transfer from an SE input buffer connected to an SE of another row not represented in the subsystem. The firing rate is set to $\lambda = -\ln(1-\xi)$, where $\xi$ is determined by the second input buffer of the SE (which is connected to an SE existing in the row model) during the previous iteration. $\xi$ is the ratio of the expected value that there is a new packet in the buffer to the expected value that there is a new packet or the buffer is empty:

$$\xi = \frac{P\{\text{NewPacket}\}}{P\{\text{NewPacket}\} + P\{\text{BufferEmpty}\}}. \tag{6.2}$$

The firing rate of the single deterministic transition, which represents the clock rate, is set to 1.

The removal of a packet in an SE output connected to an SE not represented in the row model (i.e., destined to the input of an SE not represented at the next stage) is also performed by a transition of exponentially distributed firing time. The packet passes the next stage after the expected delay time of the next stage. The delay time $d_{ns}$ of the next stage is determined by the previous iteration. The firing rate of the transition that removes the packet is set to

$$\mu = -\ln(1 - \frac{1}{d_{ns}}).\tag{6.3}$$

Although the model size is reduced by profiting from decomposition and symmetry, the state space is too large to analyze the models. Therefore, simulation must be used. The iteration is performed in two precision steps. To achieve the results of Sect. 6.5, a termination criterion of a confidence level of 80% and an estimated precision of 10% has been set for the first precision step. For the second precision step, a confidence level of 95% and a precision of 2% have been used. The iteration stops if the difference in the iteration variable of two succeeding iterations remains less than a certain bound. In this iteration, a relative difference of 5% for the first precision step is applied, and 1% for the second one.

With these constraints, the simulation run time could be reduced to about 20 hours. This still large run time is caused by the Petri net overhead. To accelerate simulation, this overhead must be eliminated by establishing a simulation model particularly adapted to the system in question (see Sect. 6.2).

### 6.1.3 Multicast Probabilities

In this section, the multicast probabilities $\omega_{mult}(k)$ are derived to model the multicast traffic under investigation. $\omega_{mult}(k)$ describes the probability that a packet at an SE input at stage $k$ is destined to $mult$ of the SE's outputs. The multicast probabilities are needed to represent the network traffic in the models, for instance, the previously established Petri net.

Let $a(i)$ denote the probability that a packet arriving at a MIN input is directed to a destination set of $i$ MIN outputs ($1 \leq i \leq N$). This means that all $a(i)$ represent the given global distribution in space of the network multicast traffic.

For the following derivation, the definition of a subnetwork is needed. A subnetwork is a part of a MIN that is itself a MIN with one less stage. For instance, in Fig. 6.7, the three-stage MIN includes two subnetworks of two stages. In general, an $n$-stage MIN ($c^n \times c^n$ MIN) with $c \times c$ SEs includes $c$ subnetworks of $n - 1$ stages ($c^{n-1} \times c^{n-1}$ MINs). Each of these subnetworks again consists of $c$ subnetworks, which are of size $c^{n-2} \times c^{n-2}$, and so on [207].

If a packet arrives at an input of a network (or subnetwork), it takes an output of the SE at the first stage to reach the following subnetwork if at

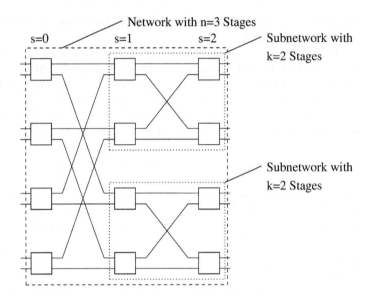

**Fig. 6.7.** 4×4 subnetworks of an 8×8 MIN

least one output of this subnetwork is the packet's destination. For instance, in Fig. 2.14 of Sect. 2.3.10, a packet arriving at the upper left $c \times c$ SE is sent via the first (upper) SE output to the upper subnetwork if at least one of the subnetwork outputs is a destination of the packet. Maybe the packet is destined to network outputs that are located in exactly *mult* subnetworks. Then, a multicast to *mult* outputs of the SE in the first network stage occurs.

In the following, the multicast probabilities $\omega_{mult}(k)$ of stage $k$ of a $n$-stage MIN are determined by calculating the multicast probabilities of the first stage of the $s$-stage subnetwork starting at the investigated stage $k$ of the $n$-stage network. Because $k$ is numbered from 0 to $n - 1$, the relation $s = n - k$ holds.

Now, consider a $c^s \times c^s$ (sub)network and a packet arriving at a (sub)network input, which is destined to $i$ (sub)network outputs. Then, the number of combinations $b_{tot}(s, i)$ in which the $i$ destinations can be spread over the $c^s$ network outputs is given by

$$b_{tot}(s, i) = \binom{c^s}{i}. \tag{6.4}$$

The number of combinations in which the $i$ destinations are spread over only the outputs that are located in exactly *mult* of the $c$ subnetworks consisting of $s - 1$ stages is denoted by $b_{mult}(s, i)$. For instance, $b_1(s, i)$ is the number of combinations in which all $i$ destinations are located in only one of the subnetworks. There are $\binom{c^{s-1}}{i}$ possibilities to distribute all destinations in a subnetwork of size $c^{s-1} \times c^{s-1}$. Because each network consists of $c$ subnetworks out of which we can choose one to distribute to all destinations, the number of combinations in which all $i$ destinations are located in only one of the subnetworks is

$$b_1(s, i) = c \cdot \binom{c^{s-1}}{i}. \tag{6.5}$$

Of course, this is only true for $1 \leq i \leq c^{s-1}$. Otherwise, we get $b_1(s, i) = 0$.

To determine the number of combinations in which the $i$ destinations are spread over outputs that are located in exactly $mult > 1$ of the $c$ subnetworks, the number of eligible subnetwork outputs is counted first. $mult$ subnetworks, with each of them consisting of $c^{s-1}$ outputs, result in $mult \cdot c^{s-1}$ eligible outputs. Spreading $i$ destinations over these outputs leads to $\binom{mult \cdot c^{s-1}}{i}$ combinations. But these combinations also include combinations where the destinations are chosen from fewer than $mult$ subnetworks; all destinations may be chosen from only one subnetwork. These combinations have to be subtracted: as defined above, $b_\ell(s, i)$ is the number of combinations where the $i$ destinations are spread over outputs that are located in exactly $\ell$ of the $c$ subnetworks $(1 \leq \ell \leq mult - 1)$, including all $\binom{c}{\ell}$ permutations to choose $\ell$ out of the $c$ subnetworks. Then, $b_\ell(s, i)/\binom{c}{\ell}$ is the number of such combinations for $\ell$ fixed subnetworks (one fixed permutation).

Out of the investigated $mult$ subnetworks, there are $\binom{mult}{\ell}$ ways to choose $\ell$ subnetworks. Eliminating these ways from all combinations found above leads to

$$b_{mult}(s, i) = \binom{c}{mult}$$
$$\cdot \left[ \binom{mult \cdot c^{s-1}}{i} - \sum_{\ell=1}^{mult-1} \binom{mult}{\ell} \frac{b_\ell(s, i)}{\binom{c}{\ell}} \right], \tag{6.6}$$

where $\binom{c}{mult}$ denotes the number of possibilities to choose $mult$ subnetworks out of $c$. Of course, Eq. (6.6) is only valid if the number of destinations $i$ is greater than or equal to $mult$ and if it is less than or equal to all investigated subnetwork outputs $mult \cdot c^{s-1}$: $mult \leq i \leq mult \cdot c^{s-1}$. Otherwise, $b_{mult}(s, i) = 0$.

The ratio $b_{mult}(s, i)/b_{tot}(s, i)$ describes the ratio of combinations that are destined to exactly $mult$ subnetworks related to all combinations. As a result, the ratio gives the probability with which a multicast to $mult$ SE outputs at the first network stage (and therefore to $mult$ subnetworks) of the $c^s \times c^s$ network occurs if a packet arriving at a network input is destined to $i$ network outputs.

The probability that a packet arriving at a $c^s \times c^s$ network's input is destined to $i$ network outputs is denoted by $a_k(i)$, and is determined later (the $c^s \times c^s$ network is a subnetwork of our investigated $c^n \times c^n$ network starting at stage $k = n - s$).

If all numbers $i$ of possible packet destinations and the probabilities $a_k(i)$ of their occurrence are considered, the multicast probabilities of the first network stage of the $c^s \times c^s$ MIN, and, therefore, the multicast probabilities of stage $k$ of the $c^n \times c^n$ MIN, are received:

$$\omega_{mult}(k) = \omega_{mult}(n-s) = \sum_{i=1}^{c^s} \frac{b_{mult}(s,i)}{b_{tot}(s,i)} \cdot a_k(i). \tag{6.7}$$

To calculate $\omega_{mult}(k)$ for all stages $k$ of the $c^n \times c^n$ network requires all $a_k(i)$ to be determined. They can be derived as follows.

At the first stage $k = 0$ of the network, $a_0(i)$ is identical to $a(i)$ of the given global network traffic pattern. With this initialization, the recursive calculation of all succeeding stages is possible: if $a_{k-1}(i)$ is the probability that a packet arriving at stage $k - 1$ is a multicast to $i$ outputs of the $c^{s+1} \times c^{s+1}$ subnetwork beginning at stage $k - 1$, the probability $a_k(j)$ that a packet arriving at the succeeding stage $k$ is a multicast to $j$ outputs of the $c^s \times c^s$ subnetwork beginning in stage $k$ is given by

$$a_k(j) = \frac{\displaystyle\sum_{i=j}^{(c-1)\cdot c^s + j} a_{k-1}(i) \frac{\binom{c^s}{j}\binom{(c-1)\cdot c^s}{i-j}}{\binom{c^{s+1}}{i}}}{\displaystyle\sum_{i=1}^{(c-1)\cdot c^s} a_{k-1}(i) \frac{\binom{c^{s+1}}{i} - \binom{(c-1)\cdot c^s}{i}}{\binom{c^{s+1}}{i}} + \sum_{i=(c-1)\cdot c^s + 1}^{c^{s+1}} a_{k-1}(i)}, \tag{6.8}$$

with $s = n - k$, $0 < k < n$, and $1 \le j \le c^s$. The numerator represents the sum of the ratios of the destination sets of the $c^{s+1} \times c^{s+1}$ network that have $j$ outputs in the $c^s \times c^s$ subnetwork under investigation as destination and the remaining $i - j$ outputs in the other $c - 1$ subnetworks that are located at the same $c^{s+1} \times c^{s+1}$ network stages. This ratio is weighted with the probability $a_{k-1}(i)$ that a packet arriving at the first stage of the $c^{s+1} \times c^{s+1}$ network is a multicast to $i$ outputs.

Before explaining the denominator, an example is presented: Figure 6.8 shows a MIN consisting of $n = 2$ stages and $2 \times 2$ SEs ($c = 2$). The network multicast traffic pattern $a(i)$ is given, and therefore, $a_0(i) = a(i)$ is known. To determine $a_1(j)$ of the subnetworks (in this case, SEs) starting at stage 1 (the last stage), Eq. (6.8) leads to ($k = 1$, $s = n - k = 1$)

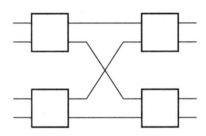

**Fig. 6.8.** Two-stage MIN consisting of $2 \times 2$ SEs

$$a_1(j) = \frac{\sum_{i=j}^{2+j} a_0(i) \frac{\binom{2}{j}\binom{2}{i-j}}{\binom{4}{i}}}{\sum_{i=1}^{2} a_0(i) \frac{\binom{4}{i}-\binom{2}{i}}{\binom{4}{i}} + \sum_{i=3}^{4} a_0(i)}. \tag{6.9}$$

For instance, $a_1(j)$ of the upper right subnetwork (SE) is to be determined. With probability $a_0(i)$, packets entering the network are directed to $i$ of the four network outputs. There exist $\binom{4}{i}$ output combinations for them. All of those combinations in which $j$ outputs are chosen from the two outputs of the upper right subnetwork ($\binom{2}{j}$) and the remaining $i - j$ are chosen from the two outputs of the other (lower right) subnetwork ($\binom{2}{i-j}$) result in traffic to $j$ outputs of the investigated (upper right) subnetwork: the numerator is established. This example is resumed later.

The limits of the numerator's sum (Eq. (6.8)) are caused by the following: the smallest considered number of outputs $i$ is $j$, because if there is a packet that is a multicast to $i$ outputs ($i < j$) in a network, the packet will be a multicast at most to $i$ outputs in a subnetwork, and the investigated number $j$ is not reached. The highest considered $i$ is $(c-1) \cdot c^s + j$, because if there is a packet that is a multicast to $i$ outputs ($i > (c-1) \cdot c^s + j$) in a network, the packet will be a multicast at least to $i - (c-1) \cdot c^s$ outputs in the investigated subnetwork (in the other subnetworks, there can be a multicast at most to all $(c-1) \cdot c^s$ outputs). In consequence, if $i < j$ or $i > (c-1) \cdot c^s + j$ there cannot be a multicast to $j$ outputs in the investigated subnetwork.

The denominator represents the destination sets that cause the packet to be sent to the investigated subnetwork. In the first sum ($1 \le i \le (c-1) \cdot c^s$), $\binom{c^{s+1}}{i}$ represents all possible destination sets with $i$ outputs as destination. In $\binom{(c-1) \cdot c^s}{i}$ cases, however, all $i$ outputs are routed to the other $c-1$ subnetworks. If $(c-1) \cdot c^s + 1 \le i \le c^{s+1}$ (second sum), the packet must always be sent to all subnetworks (including the investigated one) because of the large number of destinations. All $\binom{c^{s+1}}{i}$ cases are relevant.

In the example just presented (Eq. (6.9)), no packets reach the investigated (upper right) subnetwork if all $i$ outputs that are the destinations of the packet are located in the other (lower right) subnetwork ($\binom{2}{i}$ combinations). The remaining $\binom{4}{i} - \binom{2}{i}$ combinations cause traffic in the upper right subnetwork, and have to be considered. If $i$ is greater than the number of outputs of the other (lower right) subnetwork ($i > 2$), there must exist packet destinations in the investigated (upper right) subnetwork, and all combinations are relevant.

## 6.2 Simulation: *MINSimulate*

The simulator presented here is designed particularly to model multistage interconnection networks. Besides MINs with the banyan property as referred to in the previous section, Clos networks, bidirectional MINs, replicated MINs, multilayer MINs, and crossbars are also supported by the simulator. The simulator is named *MINSimulate* [208].

### 6.2.1 Simulator Engineering

The main goal in designing this simulator was to establish a model for faster determination of MIN performance results than possible by Petri nets. Furthermore, it also helped in the validation of mathematical model results (see Sect. 6.3). Therefore, the simulator was built in such a way that it deals in more detail with the behavior of MINs. The entire network is modeled, rather than only a single row of SEs. Packets are distinguishable, and all dependences are considered.

### Overhead

To avoid any overhead as with high-level description techniques like Petri nets, the simulation is performed by C++ code. The network is represented as a directed graph starting at the source nodes (network inputs) and ending at the destination nodes (network outputs). Packets are generated at the sources. Each packet is provided with a tag determining its destination. Due to multicasting, this tag is modeled by a vector of $N$ binary elements, each representing a network output. The elements of the desired outputs are set to "1"; all others to "0". If the packet arrives at a $c \times c$ SE, the tag is divided into $c$ subtags of equal size. Each subtag belongs to one switch output; the first (lower indices) subtag belongs to the first output, and so on. If a subtag contains at least one "1" value, a copy of the packet is sent to the corresponding output, containing the subtag as the new tag. Figure 6.9 gives an example. A part of an $8 \times 8$ MIN consisting of $2 \times 2$ SEs is shown. A packet, destined to outputs 1, 2, and 3, resulting in tag 01110000, crosses the network. At each stage, the tag is divided in the middle into two subtags ($c = 2$). Due to the existence of at least one "1" value in the subtag, a copy of the packet is sent to the corresponding SE output.

To keep the allocated memory as small as possible, only representations of the packets, referred to as containers, are routed along the network paths. These containers are replaced by the actual packets at the network outputs, allowing evaluations. Figure 6.10 gives a sketch of the simulation model.

`ContainerMultiputs` (CMs) receive the containers and store them in queues. At the first network stage, `FirstContainerMultiputs` (FCMs) additionally perform the replacement of the packets by containers.

`ContainerOutputs` (COs) send the containers to the next network stage. At the last stage, `LastContainerOutputs` (LCOs) additionally replace the containers with the corresponding packets. Each operation of a switch is controlled by its `Crossbar Manager`. The clocks perform the sequencing of the parallel actions due to single processor computer simulation.

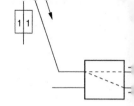

**Fig. 6.9.** Multicasting by tag

## Data Representation

As already mentioned in Sect. 4.1, packets that enter the multista
connection network are stored at a dedicated computer memory loca
includes the header of the packet and all the payload. During the
movement from buffer to buffer through the MIN, only a pointer to t
is moved from data structure to data structure representing the buff
a representation decreases the memory required as well as the simulat
It is obvious that the allocated memory is reduced due to the less
needed for a pointer, as opposed to an entire packet. The reduction
lation time results from less data to be moved during simulation. If a
data were to be moved, this process would be much more time c
than simply moving the pointers, as in *MINSimulate*.

## Confidence

Confidence level and precision of simulation results are observed by t
*Akaroa*. The simulation is stopped when the termination criteria
*Akaroa* has been developed at the University of Canterbury, New
[155].

The simulation is started by *Akaroa*. It handles the simulation ru
serves the given measures in a simulation sequence. Each result is tr
to *Akaroa* via functions in the programming language C. *Akaroa* calc
current estimated precision and confidence level. Spectral analysis, a
the method of batch means, can be applied for variance estimation. I
mination criteria given by the user are fulfilled, the simulation run is
and the final results are printed.

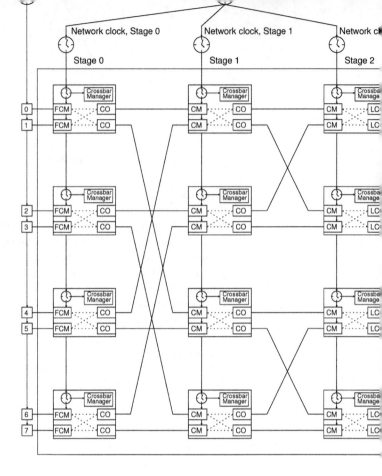

**Fig. 6.10.** Sketch of the simulation model

Intermediate results, precision, and confidence are available
simulation run via the *Akaroa* package, which includes several pr

**Parallelism and Data Recycling**

The toolkit *Akaroa*, which handles the simulation run, offers parallelism in simulation. Replications of the simulation program are started on different processors in parallel. *Akaroa* centrally collects the results of all observed measures and of all replications. Estimated precision and confidence are calculated. If termination criteria are fulfilled, the multiple replications in parallel (MRIP) are stopped. $i$ independent replications lead to an acceleration of simulation time close to $i$. Some slight losses are caused by communication with the central unit of *Akaroa* (called master).

As already described in Sect. 4.1, multiple replications lead to fault tolerance in the simulation. If a replication on a processor fails, the remaining replications still deliver results to the *Akaroa* master, and the simulation still comes to a successful end.

Data recycling is also applied in *MINSimulate*. If multiple measures of MIN are of interest, all of them are determined during a single simulation run.

If measures consist of parameters dependent on time, inefficient individual simulations for each time step can be replaced by observing all measures of all time steps in a single simulation run. *MINSimulate* in combination with *Akaroa* offers both methods: a fast method to determine all measures of all time steps in a single simulation, and individual simulations for each time step if the number of measures exceeds the amount that can be handled by the computer's main memory: in the case of a single simulation, each time step must be covered by its own measure set, resulting in a huge number of measures if many time steps are observed.

### 6.2.2 Features

The network to be evaluated is determined by the user via a graphical user interface (GUI). Figure 6.11 shows the main tab to set simulation parameters. Parameters that are not available in the currently chosen network configuration are blind out. A short sketch of the parameters and their available settings is given below. They are described in more detail in [208].

First, it can be chosen whether to simulate single crossbars or MINs. If MINs are chosen, their type must be specified: MINs with the banyan property, Clos networks, and bidirectional MINs are available. To simulate replicated or multilayer MINs, the `Banyan` item must be chosen in combination with the `Multilayer` tab. There, `constant number of layers` refers to replicated MINs.

Input buffers can be chosen as shared ones (with a minimum size and maximum size for each crossbar input, as well as the overall buffer size of a crossbar) or as non-shared ones (with the size per crossbar input). Tab `Special buffer configuration` allows individual buffer settings for each stage.

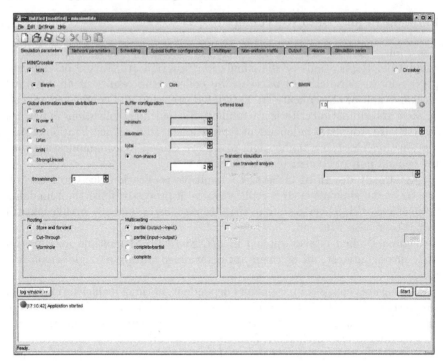

**Fig. 6.11.** Main tab of *MINSimulate*

The global address destination distribution of packets entering the network can also be varied. The most important patterns are `onl1` (only unicast; all targets are packet destinations with equal probability), `N over K` (multicast; all target combinations are packet destinations with equal probability), `Ufun` (multicast with many unicasts and many broadcasts), and `onlN` (only broadcast). If single sources are desired to produce deviating address distributions, tab `Non-uniform traffic` helps.

The parameter `Streamlength` refers to the number of packets a message consists of. It allows generating sequences of packets that are destined to the same destination node.

When choosing the routing algorithm, the following packet switching schemes are available: store-and-forward switching, (virtual) cut-through switching, and wormhole switching. Multicast in the case of wormhole switching usually suffers from deadlocks. The wormhole switching algorithm of *MINSimulate* avoids deadlocks by grouping appropriate parts of the network [225]. Wormhole switching requires dividing the packets into flits. The number of flits per packet is also a parameter for the simulation.

The kind of multicast can be set to complete multicast, partial multicast, or a two-phase version of both, where complete multicast is first applied to an SE, and, in the second phase, partial multicast is performed for all remaining SE outputs. Another parameter represents the offered load to each network input. The last parameter in the main tab determines whether to observe measures in time (terminating simulation) instead of observing the steady state. In the case of transient simulation, the number of clock cycles to simulate can be fixed in the `Transient simulation` area.

Crossbar size or MIN size are defined via tab `Network parameters` (see Fig. 6.12). If MINs are chosen, two of the three parameters of the equation

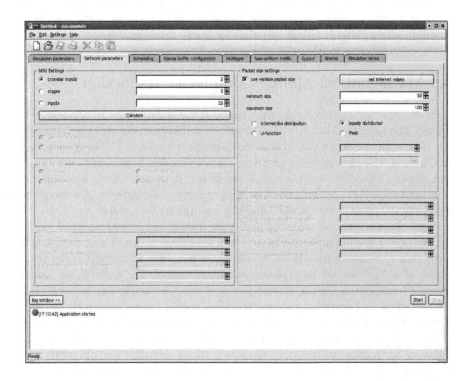

**Fig. 6.12.** Network parameter tab of *MINSimulate*

$n = \log_c N$ must be set by the user; the third is calculated by the GUI. Packets may be defined to be generated with variable size. For instance, the packet size distribution known from the Internet may be of interest. Other parameters at this tab relate to Clos networks (blind out in Fig. 6.12).

Instead of simulating a particular network configuration, a parameter can be varied to deal with parameter-dependent results. In tab `Simulation series`, the parameter to vary is chosen. A start value, an end value, and a step size determine the variation. If desired, step size can be changed once in the parameter interval.

Performance measures are chosen via tab `Output`. The most important ones are throughput, delay times, and queue lengths. A histogram of delay times within an interval is also available. Deadlines can be added to packets, and packets that exceed their deadline are then removed. In such a scenario, packet loss is a measure.

*Akaroa* parameters to determine confidence level and estimated precision of results are set in tab `Akaroa`.

Results achieved by *MINSimulate* are presented in Sect. 6.5. Compared to the 20 hours of simulation (computation) run time of Petri nets, *MINSimulate* obtains results in about four hours. However, although this heavily reduces simulation run time, it is still too high for many investigations.

## 6.3 Mathematical Model: Complexity Reduction

To overcome the high computation times of the simulations previously presented, mathematical models are now derived. This section and the following one deal with the development of a mathematical model of MINs. As outlined in Chap. 4, the development of mathematical models is generally very time consuming. Therefore, only a simple architecture of multistage interconnection networks, this one with the banyan property, is considered. A concept to generally decrease development time is presented in Sect. 6.4.

This section demonstrates the basics of establishing a mathematical model by taking the high complexity of such models into account. Thus, complexity reduction in mathematically modeling MINs is a very important issue. If all states of a MIN would be represented in the model, the state space would become too large. Even the state space of a small network, like an 8×8 MIN consisting of 2×2 switching elements (SEs), would exceed ordinary memory size: assuming that each SE input can adapt only two states, three stages of eight SE inputs each would result in $2^{24}$ network states! But, in the case of buffered SE inputs, many more than two states describe the SE input and its buffer. The buffer size and the number of different packet types determine the number of distinguishable states of an SE input. For instance, ten packet types and a buffer of the size of ten packets would lead to $10^{10}$ states for each SE input (considering only a completely filled buffer) and $(10^{10})^{24} = 10^{240}$ states could be distinguished for the entire MIN. Thus, complexity reduction is essential.

### 6.3.1 Symmetries

In this example, profiting from the network symmetries reduces network complexity. A short sketch of this idea was presented in Sect. 4.2.2. Symmetric network architecture combined with symmetric (uniform) network traffic leads to a similar behavior of network subsystems. A representation of the entire network (Fig. 4.6(a)) can be replaced by a single switching element row of input-output pairs (Fig. 4.6(c)). As described in Sect. 4.2.2, this subsystem cyclically depends on itself because inputs (mean values of throughput, delay, and queue length) to this row that originate from another row are also represented by this single row. The uniform traffic precondition is guaranteed by some assumptions, which are similar to those of the Petri net model of Sect. 6.1:

- The traffic load to all inputs of the network is equal.
- Packet destinations are uniformly distributed. This means that every output of the network is with equal probability one of the destinations of a packet.
- Conflicts between packets for network resources are randomly solved with equal probabilities.
- The switching of all elements is synchronously performed with an internal clock cycle.
- Routing is performed in a pipeline manner. This means that the routing process occurs in every stage in parallel.

Furthermore, some additional assumptions simplify the dependences and also reduce the state space:

- All packets have the same size.
- Packets are immediately removed from their final destinations after arrival.
- The destinations of succeeding packets are independent of each other.

These assumptions allow establishing a mathematical model to determine the performance of $N \times N$ MINs consisting of $2 \times 2$ SEs with $n = \log_2 N$ stages. At each stage $k$ $(0 \leq k \leq n - 1)$, there is a FIFO buffer of size $m_{max}(k)$ in front of each SE input (the Petri net model previously presented dealt only with a FIFO buffer to store a single packet). The packets are routed by store-and-forward switching from one stage to its succeeding one by the backpressure mechanism. The derivation for cut-through switching is similar to the following one except some minor differences. It can be found in [209] (see also Sect. 6.4.3).

Discrete time Markov chains (DTMCs) are chosen as modeling technique. The model is an extension of Jenq's model I [85]. However, the following model is also able to deal with multicast traffic. A detailed description of the model can be found in [203].

Due to the uniform traffic, the state of each network stage is represented by the state of a single SE input buffer at that stage. With regard to the first position in the buffer, seven states can be distinguished.

State $n$ represents a normal packet (that means a non-broadcast packet) in the first buffer position (not including the states $nb$, $fb$, and $nbfb$). The probability that a buffer at stage $k$ at time $t$ is in this state is given by $\pi_n(k, t)$.

State $nb$ represents a normal packet that was blocked because the destination buffer was full or made full by the packet in the other SE input sent to it (state $nb$ does not include state $nbfb$). Such a packet is called a blocked normal packet. The probability that a buffer at stage $k$ at time $t$ is in this state is given by $\pi_{nb}(k, t)$.

State $b$ represents a broadcast packet in the first buffer position (not including state $bb$). The probability that a buffer at stage $k$ at time $t$ is in this state is given by $\pi_b(k, t)$.

State $bb$ represents a broadcast packet that was blocked because both destination buffers were full or made full by the packet in the other SE input sent to these destination buffers. Such a packet is called a blocked broadcast packet. The probability that a buffer at stage $k$ at time $t$ is in this state is given by $\pi_{bb}(k, t)$.

State $fb$ represents the case where there have been broadcast packets in the first position of both buffers of an SE. These packets have been in conflict with each other, and the conflict has been resolved in such a way that only one copy of each was sent. The destination buffer examined has not been made full by one of the copies. For instance, the broadcast packet in the upper input wins the conflict for the upper output and the broadcast packet in the lower input wins the conflict for the lower output (dashed lines in Fig. 6.13). Then, each packet sends a copy to its reserved output. In the upper input, there remains a packet destined to the lower output, and in the lower input, there remains a packet destined to the upper output (solid line in Fig. 6.13). Both

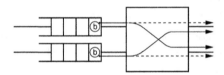

**Fig. 6.13.** Remaining packets change to state $fb$

buffers are in the state $fb$ because it is known that there is no conflict between their packets in the next clock cycle. The probability that a buffer at stage $k$ at time $t$ is in this state is given by $\pi_{fb}(k, t)$.

State $nbfb$ represents the case where there have been broadcast packets in the first position of both buffers of an SE. These packets have been in conflict with each other, and the conflict was resolved in such a way that only a copy of each was sent. The destination buffer examined has been made full by one of the copies (the difference with state $fb$.). The probability that a buffer at stage $k$ at time $t$ is in this state is given by $\pi_{nbfb}(k, t)$.

State $0$ represents the case where the first buffer position is empty. This means that the entire buffer is empty. The probability that a buffer at stage $k$ at time $t$ is in this state is given by $\pi_0(k, t)$.

Successively determining the state probabilities for each clock cycle (represented by the time $t$) dependent on the previous one finally leads to steady-state network performance. To calculate the state of a buffer in the next clock cycle, it is important to know whether a packet left the buffer and whether a new packet arrived at it.

## Probability of Sending Packets

In the following, the probability of sending a packet to the next stage is derived. For this analysis, all previously introduced states of a packet in the first buffer position have to be independently considered. The probability that a packet in the first buffer position of stage $k$ ($0 \leq k \leq n-1$) will be forwarded to the next stage at time $t$ is given by $r(k, t)$. If the first buffer position contains a normal packet (Fig. 6.14), the buffer is in state $n$. The probability that this packet will be forwarded to the next stage is denoted by $r_n(k, t)$. To calculate

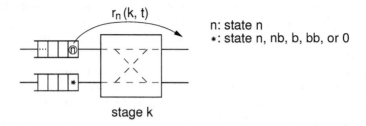

**Fig. 6.14.** Sending a normal packet

this probability, all possible behaviors of the two input buffers of the following stage $k + 1$ have to be considered. We distinguish behaviors $\alpha$, $\bar{\alpha}$, $\beta$, and $\bar{\beta}$. Behavior $\alpha$ is defined to describe the case where it is not known whether the buffer at stage $k + 1$ is full. It is guaranteed, however, that the packet will not be blocked. This can be the case if either the buffer at stage $k + 1$ is not full or the buffer is full but one packet is leaving. Let the probability for this behavior be $\alpha(k + 1, t)$. Correspondingly, behavior $\bar{\alpha}$ is defined to represent the case where the buffer is full and no packet will leave, so that the packet in stage $k$ will be blocked. The probability for this behavior is $1 - \alpha(k + 1, t)$.

In contrast, behavior $\beta$ means that it is known that the buffer at stage $k+1$ is full, for example, because the buffer of the SE in front of it has a blocked broadcast packet in the first position (state $bb$), but buffer space will become available. The probability that buffer space becomes available in such a case is given by $\beta(k + 1, t)$. Correspondingly, behavior $\bar{\beta}$ is defined to represent the case where no buffer space will become available. The probability for this behavior is $1 - \beta(k + 1, t)$.

The probability $r_n(k, t)$ that a normal packet in the first buffer position will be sent to the next stage is

$$r_n(k, t) = r_{n\alpha\alpha}(k, t) + r_{n\alpha\beta}(k, t) + r_{n\beta\beta}(k, t) + r_{n\alpha\bar{\alpha}}(k, t)$$
$$+ r_{n\alpha\bar{\beta}}(k, t) + r_{n\bar{\alpha}\beta}(k, t) + r_{n\beta\bar{\beta}}(k, t), \tag{6.10}$$

where each term of the sum represents a behavior of the buffers at stage $k+1$: thus, $r_{n\alpha\alpha}(k,t)$ is the probability that the normal packet will be sent if both of the following buffers show behavior $\alpha$.

The two buffers of stage $k+1$ are not distinguished. $r_{n\alpha\beta}(k,t)$ represents the case where one of the two buffers shows behavior $\alpha$ and the other buffer behavior $\beta$. No definition $r_{n\beta\alpha}(k,t)$ is required.

To send a normal packet out of stage $k$, the destination buffer of this packet must be or become available. This means that at least one of the following buffers must show behavior $\alpha$ or $\beta$. If the next stage buffers show behavior $\bar{\alpha}\bar{\alpha}$, $\bar{\alpha}\bar{\beta}$, or $\bar{\beta}\bar{\beta}$, the packet will be blocked.

In the following, the sending probabilities of a normal packet for all possible input buffer behaviors of stage $k+1$ are determined. It is started with the probability $r_{n\alpha\alpha}(k,t)$ that the normal packet at stage $k$ will be sent and the second buffer of this SE at stage $k$ is in such a state that both buffers of stage $k+1$ following the SE considered show behavior $\alpha$. Because these buffers are independent of each other, they show this behavior with probability $\alpha^2(k+1,t)$.

One input buffer of the SE at stage $k$ is in state $n$, as assumed above. Then, the second input buffer must be in state 0, state $n$, or state $b$. If it were to be in one of the other states, at least one of the buffers at stage $k$ would have to show behavior $\beta$ or $\bar{\beta}$. But this behavior is not considered yet. If the second input buffer is in state 0 (probability $\pi_0(k,t)$), there is no conflict, and the normal packet considered will be forwarded with the probability 1. If the buffer is in state $n$ (probability $\pi_n(k,t)$), the normal packet considered will be forwarded if different SE outputs are the destination (probability 0.5) or the same SE output is the destination (probability 0.5) and the considered packet wins the conflict (probability 0.5). If the buffer is in state $b$ (probability $\pi_b(k,t)$), the normal packet considered will be forwarded if it wins the unavoidable conflict (probability 0.5).

Because of the normal packet in one input buffer, the state space of the second one is reduced: it cannot be in state $fb$ or $nbfb$. These two states are only possible if both input buffers are in one of these states. To adapt the probabilities $\pi_0(k,t)$, $\pi_n(k,t)$, and $\pi_b(k,t)$, which are defined for the complete state space, to the reduced state space, we have to use a normalization term, $\psi_{\overline{fb}}(k,t)$, which is determined later. So, we get

$$r_{n\alpha\alpha}(k,t) = \alpha^2(k+1,t) \ (\pi_0(k,t) \cdot 1 + \pi_n(k,t) \cdot (0.5 + 0.5 \cdot 0.5)$$
$$+ \pi_b(k,t) \cdot 0.5) \cdot \frac{1}{\psi_{\overline{fb}}(k,t)}. \tag{6.11}$$

All other probabilities of the sum in Eq. (6.10) are derived in the same manner. They are listed below to finally give the entire system of equations:

$$r_{n\alpha\beta}(k,t) = \alpha(k+1,t)\beta(k+1,t)\pi_{nb}(k,t) \cdot (0.5 + 0.5 \cdot 0.5)$$
$$\cdot \frac{1}{\psi_{\overline{fb}}(k,t)}, \tag{6.12}$$

$$r_{n\beta\beta}(k,t) = \beta^2(k+1,t)\pi_{bb}(k,t) \cdot 0.5 \cdot \frac{1}{\psi_{\overline{fb}}(k,t)}, \tag{6.13}$$

$$r_{n\alpha\bar{\alpha}}(k,t) = \alpha(k+1,t)(1-\alpha(k+1,t)) \cdot 2 \cdot 0.5$$
$$\cdot (\pi_0(k,t) + \pi_n(k,t) \cdot (0.5 + 0.5 \cdot 0.5) + \pi_b(k,t) \cdot 0.5)$$
$$\cdot \frac{1}{\psi_{\overline{fb}}(k,t)}, \tag{6.14}$$

$$r_{n\alpha\bar{\beta}}(k,t) = \alpha(k+1,t)(1-\beta(k+1,t)) \cdot 2 \cdot 0.5$$
$$\cdot \pi_{nb}(k,t) \cdot 0.5 \cdot \frac{1}{\psi_{\overline{fb}}(k,t)}, \tag{6.15}$$

$$r_{n\bar{\alpha}\beta}(k,t) = \beta(k+1,t)(1-\alpha(k+1,t)) \cdot 2 \cdot 0.5$$
$$\cdot \pi_{nb}(k,t) \cdot 0.5 \cdot 0.5 \cdot \frac{1}{\psi_{\overline{fb}}(k,t)}, \text{ and} \tag{6.16}$$

$$r_{n\beta\bar{\beta}}(k,t) = \beta(k+1,t)(1-\beta(k+1,t)) \cdot 2 \cdot 0.5$$
$$\cdot \pi_{bb}(k,t) \cdot 0.5 \cdot \frac{1}{\psi_{\overline{fb}}(k,t)}. \tag{6.17}$$

A detailed description of the equations (and all following ones of this section) can be found in [203].

If the first buffer position of stage $k$ contains a blocked normal packet, the buffer is in state $nb$. The probability that this packet will be forwarded to the next stage is given by $r_{nb}(k,t)$. It is calculated with similar considerations as $r_n(k,t)$:

$$r_{nb}(k,t) = r_{nb\alpha\beta}(k,t) + r_{nb\beta\beta}(k,t) + r_{nb\bar{\alpha}\beta}(k,t) + r_{nb\beta\bar{\beta}}(k,t), \tag{6.18}$$

where each term of the sum represents a behavior of the buffers at stage $k+1$ in which a transmission is possible:

$$r_{nb\alpha\beta}(k,t) = \alpha(k+1,t)\beta(k+1,t)$$
$$\cdot (\pi_0(k,t) + \pi_n(k,t) \cdot 0.75 + \pi_b(k,t) \cdot 0.5 + \pi_{nb}(k,t) \cdot 0.25)$$
$$\cdot \frac{1}{\psi_{\overline{fb}}(k,t)}, \tag{6.19}$$

$$r_{nb\beta\beta}(k,t) = \beta^2(k+1,t)(\pi_{nb}(k,t) \cdot 0.5 + \pi_{bb}(k,t) \cdot 0.5) \cdot \frac{1}{\psi_{\overline{fb}}(k,t)}, \tag{6.20}$$

$$r_{nb\bar{\alpha}\beta}(k,t) = (1 - \alpha(k+1,t))\beta(k+1,t) \cdot 2 \cdot 0.5$$
$$\cdot (\pi_0(k,t) + \pi_n(k,t) \cdot 0.75 + \pi_b(k,t) \cdot 0.5 + \pi_{nb}(k,t) \cdot 0.25)$$
$$\cdot \frac{1}{\psi_{\overline{fb}}(k,t)}, \text{ and} \tag{6.21}$$

$$r_{nb\beta\bar{\beta}}(k,t) = \beta(k+1,t)(1 - \beta(k+1,t)) \cdot 2 \cdot 0.5$$
$$\cdot (\pi_{nb}(k,t) \cdot 0.5 + \pi_{bb}(k,t) \cdot 0.5) \frac{1}{\psi_{\overline{fb}}(k,t)}. \tag{6.22}$$

If the first buffer position at stage $k$ contains a broadcast packet, the buffer is in state $b$. The probability that this packet will be *completely* forwarded to the next stage is called $r_b(k,t)$;

$$r_b(k,t) = r_{b\alpha\alpha}(k,t) + r_{b\alpha\beta}(k,t) + r_{b\beta\beta}(k,t), \tag{6.23}$$

with

$$r_{b\alpha\alpha}(k,t) = \alpha^2(k+1,t)\,(\pi_0(k,t) + \pi_n(k,t) \cdot 0.5 + \pi_b(k,t) \cdot 0.25)$$
$$\cdot \frac{1}{\psi_{\overline{fb}}(k,t)}, \tag{6.24}$$

$$r_{b\alpha\beta}(k,t) = \alpha(k+1,t)\beta(k+1,t)\pi_{nb}(k,t) \cdot 0.5 \cdot \frac{1}{\psi_{\overline{fb}}(k,t)}, \text{ and} \tag{6.25}$$

$$r_{b\beta\beta}(k,t) = \beta^2(k+1,t)\pi_{bb}(k,t) \cdot 0.25 \cdot \frac{1}{\psi_{\overline{fb}}(k,t)}. \tag{6.26}$$

If the broadcast packet is not completely transferred to the next stage, for example, because only one destination buffer is available, there could be a chance to transfer a copy to one of the destination buffers. The probability that only one copy of a broadcast packet in the first buffer position of stage $k$ will be forwarded to the next stage and that state $fb$ or $nbfb$ does not result at stage $k$ is called $r_{pb}(k,t)$:

$$r_{pb}(k,t) = r_{pb\alpha\alpha}(k,t) + r_{pb\alpha\beta}(k,t) + r_{pb\alpha\bar{\alpha}}(k,t) + r_{pb\alpha\bar{\beta}}(k,t)$$
$$+ r_{pb\bar{\alpha}\beta}(k,t) + r_{pb\beta\bar{\beta}}(k,t). \tag{6.27}$$

The cases in which the state $fb$ or $nbfb$ result at stage $k$ after transmission are not considered yet. The terms of the sum in Eq. (6.27) are given by

$$r_{pb\alpha\alpha}(k,t) = \alpha^2(k+1,t)\pi_n(k,t) \cdot 0.5 \cdot \frac{1}{\psi_{\overline{fb}}(k,t)}, \tag{6.28}$$

$$r_{pb\alpha\beta}(k,t) = \alpha(k+1,t)\beta(k+1,t)\pi_{nb}(k,t) \cdot 0.5 \cdot \frac{1}{\psi_{\overline{fb}}(k,t)}, \tag{6.29}$$

$$r_{pb\alpha\bar{\alpha}}(k,t) = \alpha(k+1,t)(1-\alpha(k+1,t))\cdot 2$$
$$\cdot(\pi_0(k,t)+\pi_n(k,t)\cdot 0.75+\pi_b(k,t)\cdot 0.5)\cdot\frac{1}{\psi_{\overline{fb}}(k,t)}, \quad (6.30)$$

$$r_{pb\alpha\bar{\beta}}(k,t) = \alpha(k+1,t)(1-\beta(k+1,t))\cdot 2$$
$$\cdot\pi_{nb}(k,t)\cdot 0.5\cdot\frac{1}{\psi_{\overline{fb}}(k,t)}, \quad (6.31)$$

$$r_{pb\bar{\alpha}\beta}(k,t) = (1-\alpha(k+1,t))\beta(k+1,t)\cdot 2$$
$$\cdot\pi_{nb}(k,t)\cdot 0.25\cdot\frac{1}{\psi_{\overline{fb}}(k,t)}, \text{ and} \quad (6.32)$$

$$r_{pb\beta\bar{\beta}}(k,t) = \beta(k+1,t)(1-\beta(k+1,t))\cdot 2$$
$$\cdot\pi_{bb}(k,t)\cdot 0.5\cdot\frac{1}{\psi_{\overline{fb}}(k,t)}. \quad (6.33)$$

Now, the cases in which a transmission results in the states $fb$ or $nbfb$ at stage $k$ are considered. This is only possible if the other input buffer is in state $b$ or $bb$.

The probability that only one copy of a broadcast packet that is in the first buffer position at stage $k$ and is in conflict with another broadcast packet (state $b$ of the second input buffer) will be forwarded to the next stage and state $fb$ or $nbfb$ results at stage $k$ is given by $r_{pbfb}(k,t)$:

$$r_{pbfb}(k,t) = r_{pbfb\alpha\alpha}(k,t) = \alpha^2(k+1,t)\pi_b(k,t)\cdot 0.5\cdot\frac{1}{\psi_{\overline{fb}}(k,t)}. \quad (6.34)$$

The probability that only one copy of a broadcast packet that is in the first buffer position at stage $k$ and is in conflict with a blocked broadcast packet (state $bb$ of the second input buffer) will be forwarded to the next stage and state $fb$ or $nbfb$ results at stage $k$ is given by $r_{pbfbb}(k,t)$:

$$r_{pbfbb}(k,t) = r_{pbfbb\beta\beta}(k,t) = \beta^2(k+1,t)\pi_{bb}(k,t)\cdot 0.5\cdot\frac{1}{\psi_{\overline{fb}}(k,t)}. \quad (6.35)$$

If the first buffer position at stage $k$ contains a blocked broadcast packet, the buffer is in state $bb$. The probability that this packet will be *completely* forwarded to the next stage is given by $r_{bb}(k,t)$:

$$r_{bb}(k,t) = r_{bb\beta\beta}(k,t) = \beta^2(k+1,t)\,(\pi_0(k,t)+\pi_n(k,t)\cdot 0.5$$
$$+\pi_b(k,t)\cdot 0.25+\pi_{nb}(k,t)\cdot 0.5$$
$$+\pi_{bb}(k,t)\cdot 0.25)\cdot\frac{1}{\psi_{\overline{fb}}(k,t)}. \quad (6.36)$$

If the blocked broadcast packet is not completely transferred to the next stage, for example, because only one destination buffer is available, there could be a chance to transfer a copy to one of the destination buffers. The probability

that only one copy of a blocked broadcast packet that is in the first buffer
position at stage $k$ will be forwarded to the next stage and state $fb$ or $nbfb$
does not result at stage $k$ is given by $r_{pbb}(k,t)$:

$$r_{pbb}(k,t) = r_{pbb\beta\beta}(k,t) + r_{pbb\beta\bar{\beta}}(k,t). \qquad (6.37)$$

The cases in which the state $fb$ or $nbfb$ results at stage $k$ after transmission
are not considered yet. The terms of the sum in Eq. (6.37) are given by

$$r_{pbb\beta\beta}(k,t) = \beta^2(k+1,t)\,(\pi_n(k,t) \cdot 0.5 + \pi_{nb}(k,t) \cdot 0.5)$$
$$\cdot \frac{1}{\psi_{\overline{fb}}(k,t)} \quad \text{and} \qquad (6.38)$$
$$r_{pbb\beta\bar{\beta}}(k,t) = \beta(k+1,t)(1 - \beta(k+1,t)) \cdot 2$$
$$\cdot (\pi_0(k,t) + \pi_n(k,t) \cdot 0.75 + \pi_{nb}(k,t) \cdot 0.75$$
$$+ \pi_b(k,t) \cdot 0.5 + \pi_{bb}(k,t) \cdot 0.5) \cdot \frac{1}{\psi_{\overline{fb}}(k,t)}. \qquad (6.39)$$

Now, the cases in which a transmission results in the states $fb$ or $nbfb$ at
stage $k$ are considered. This is only possible if the other input buffer is in
state $b$ or $bb$.

The probability that only one copy of a blocked broadcast packet that is
in the first buffer position at stage $k$ and is in conflict with another broadcast
or blocked broadcast packet (state $b$ or $bb$ of the second input buffer) will be
forwarded to the next stage and state $fb$ or $nbfb$ results at stage $k$ is given
by $r_{pbbfb}(k,t)$:

$$r_{pbbfb}(k,t) = r_{pbbfb\beta\beta}(k,t)$$
$$= \beta^2(k+1,t)\,(\pi_b(k,t) \cdot 0.5 + \pi_{bb}(k,t) \cdot 0.5)$$
$$\cdot \frac{1}{\psi_{\overline{fb}}(k,t)}. \qquad (6.40)$$

The probability that a packet in the first position of a buffer in state $fb$
at stage $k$ will be forwarded to the next stage is called $r_{fb}(k,t)$. Because the
destination buffer is available (otherwise, the buffer would be in state $nbfb$)
and there is no conflict with the packet in the buffer of the other SE input
(see definition of state $fb$), the examined packet will leave the buffer in any
case:

$$r_{fb}(k,t) = 1. \qquad (6.41)$$

Although $r_{fb}(k,t)$ is a constant, this probability is defined to get a continuous
scheme.

The probability that a packet in the first position of a buffer in state $nbfb$
at stage $k$ will be forwarded to the next stage is called $r_{nbfb}(k,t)$:

$$r_{nbfb}(k,t) = r_{nbfb\alpha\beta}(k,t) + r_{nbfb\beta\beta}(k,t) + r_{nbfb\bar{\alpha}\beta}(k,t) + r_{nbfb\beta\bar{\beta}}(k,t). \qquad (6.42)$$

Because the probability $\pi_{fb}(k,t)$ is defined for the entire state space, each following term is adapted to the reduced state space of $fb$ and $nbfb$. The terms of the sum in Eq. (6.42) result in

$$r_{nbfb\alpha\beta}(k,t) = \alpha(k+1,t)\beta(k+1,t)\pi_{fb}(k,t) \cdot \frac{1}{\psi_{fb}(k,t)}, \qquad (6.43)$$

$$r_{nbfb\beta\beta}(k,t) = \beta^2(k+1,t)\pi_{nbfb}(k,t) \cdot \frac{1}{\psi_{fb}(k,t)}, \qquad (6.44)$$

$$r_{nbfb\bar{\alpha}\beta}(k,t) = (1 - \alpha(k+1,t))\beta(k+1,t) \cdot 2$$
$$\cdot \pi_{fb}(k,t) \cdot 0.5 \cdot \frac{1}{\psi_{fb}(k,t)}, \text{ and} \qquad (6.45)$$

$$r_{nbfb\beta\bar{\beta}}(k,t) = \beta(k+1,t)(1 - \beta(k+1,t)) \cdot 2$$
$$\cdot \pi_{nbfb}(k,t) \cdot 0.5 \cdot \frac{1}{\psi_{fb}(k,t)}. \qquad (6.46)$$

### Normalization and Available Destination Buffers

To normalize to the possible states of the second buffer of the SE due to the reduced state space, two terms are used:

$$\psi_{fb}(k,t) = \pi_{fb}(k,t) + \pi_{nbfb}(k,t) \quad \text{and} \qquad (6.47)$$
$$\psi_{\overline{fb}}(k,t) = \pi_0(k,t) + \pi_n(k,t) + \pi_b(k,t) + \pi_{nb}(k,t) + \pi_{bb}(k,t). \qquad (6.48)$$

To calculate the probability $\beta(k,t)$ that a packet will leave a non-empty buffer, the sum of each possible buffer state and the probability of forwarding the packet in the first buffer position as determined before must be considered. Because a non-empty buffer is examined, the probabilities must be adapted to the reduced state space with a normalization term $\psi_{\bar{0}}(k,t)$:

$$\beta(k,t) = \frac{1}{\psi_{\bar{0}}(k,t)} \; (\pi_n(k,t)r_n(k,t) + \pi_{nb}(k,t)r_{nb}(k,t)$$
$$+\pi_b(k,t)r_b(k,t) + \pi_{bb}(k,t)r_{bb}(k,t)$$
$$+\pi_{fb}(k,t)r_{fb}(k,t) + \pi_{nbfb}(k,t)r_{nbfb}(k,t)). \qquad (6.49)$$

The normalization term includes all states except the state 0 (because the buffer is not empty):

$$\psi_{\bar{0}}(k,t) = \pi_n(k,t) + \pi_{nb}(k,t) + \pi_b(k,t)$$
$$+\pi_{bb}(k,t) + \pi_{fb}(k,t) + \pi_{nbfb}(k,t). \qquad (6.50)$$

To calculate the probability $\alpha(k,t)$ that a buffer space is available, the probability $\pi_m(k,t)$ that the buffer at stage $k$ contains $m$ packets is defined. This probability introduces an additional state space. It is later referred to this probability and state space. $\pi_{m_{max}(k)}(k,t)$ is the probability that the buffer is full. Buffer space is available if the buffer is not full or if the buffer is full and a packet will leave:

$$\alpha(k,t) = (1 - \pi_{m_{max}(k)}(k,t)) + \pi_{m_{max}(k)}(k,t)\beta(k,t). \qquad (6.51)$$

**Probability of Receiving Packets**

Now, the probability of receiving a packet from the previous stage is derived. The probability that a new packet will be sent into the input buffer at stage $k + 1$ $(0 \leq k \leq n - 1)$ at time $t$ is given by $q_I(k + 1, t)$. To determine this probability, all state combinations of the both input buffers in the previous stage $k$ (except state combination 0,0) have to be independently considered. State combination 0,0 (both input buffers at stage $k$ are empty) results in $q_{I0,0} = 0$, because there is no packet to be received by stage $k + 1$. The probability is

$$\begin{aligned} q_I(k + 1, t) = \; & q_{I0,n}(k + 1, t) + q_{I0,nb}(k + 1, t) + q_{I0,b}(k + 1, t) \\ & + q_{I0,bb}(k + 1, t) + q_{In,n}(k + 1, t) + q_{In,nb}(k + 1, t) \\ & + q_{In,b}(k + 1, t) + q_{In,bb}(k + 1, t) + q_{Inb,nb}(k + 1, t) \\ & + q_{Inb,b}(k + 1, t) + q_{Inb,bb}(k + 1, t) + q_{Ib,b}(k + 1, t) \\ & + q_{Ib,bb}(k + 1, t) + q_{Ibb,bb}(k + 1, t) + q_{Ifb,fb}(k + 1, t) \\ & + q_{Ifb,nbfb}(k + 1, t) + q_{Inbfb,nbfb}(k + 1, t). \end{aligned} \tag{6.52}$$

For instance, $q_{I0,n}(k + 1, t)$ is the probability (Fig. 6.15) that a new packet

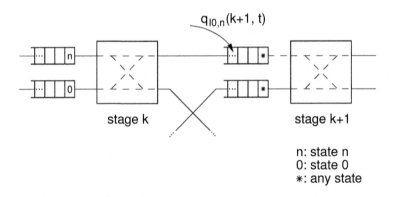

**Fig. 6.15.** Receiving packets if previous buffers are in state 0 and $n$

will be sent to the input buffer at stage $k + 1$ at time $t$ if one input buffer at stage $k$ is in state 0 (empty), and one input buffer is in state $n$ (the first buffer position contains a normal packet). Because there are two possibilities of how the states can be spread over the two inputs, this combination has probability $2 \cdot \pi_0(k, t)\pi_n(k, t)$. The reduced state space of the second input buffer, if the other is in state 0, results in the normalization term $\psi_{\overline{fb}}(k, t)$. If both input buffers at the examined stage $k + 1$ are available, the normal packet is sent in any case (probability 1) to the input buffers, resulting in the probability $\alpha^2(k + 1, t) \cdot 1$. If one input buffer of the examined stage $k + 1$ is available and the other is not available (i.e., it is full), there are two possibilities which of the two buffers is full. The normal packet is sent to the available input buffer if it is the destination of the packet (probability 0.5), resulting in the probability $\alpha(k+1,t)(1-\alpha(k+1,t)) \cdot 2 \cdot 0.5$. The sum of the last two terms results in $\alpha(k+1,t)$. Because, so far, the entire 2×2 SE with its two outputs was considered, a normalization to one output is required (factor 0.5):

$$q_{I0,n}(k+1,t) = 2 \cdot \pi_0(k,t)\frac{\pi_n(k,t)}{\psi_{\overline{fb}}(k,t)} \cdot \alpha(k+1,t) \cdot 0.5. \qquad (6.53)$$

All other terms are derived similarly:

$$q_{I0,nb}(k+1,t) = 2 \cdot \pi_0(k,t)\frac{\pi_{nb}(k,t)}{\psi_{\overline{fb}}(k,t)} \cdot \beta(k+1,t) \cdot 0.5, \qquad (6.54)$$

$$q_{I0,b}(k+1,t) = 2 \cdot \pi_0(k,t)\frac{\pi_b(k,t)}{\psi_{\overline{fb}}(k,t)} \cdot 2\alpha(k+1,t) \cdot 0.5, \qquad (6.55)$$

$$q_{I0,bb}(k+1,t) = 2 \cdot \pi_0(k,t)\frac{\pi_{bb}(k+1,t)}{\psi_{\overline{fb}}(k,t)} \cdot 2\beta(k+1,t) \cdot 0.5, \qquad (6.56)$$

$$q_{In,n}(k+1,t) = \pi_n(k,t)\frac{\pi_n(k,t)}{\psi_{\overline{fb}}(k,t)} \cdot 1.5 \cdot \alpha(k+1,t) \cdot 0.5, \qquad (6.57)$$

$$q_{In,nb}(k+1,t) = 2 \cdot \pi_n(k,t)\frac{\pi_{nb}(k,t)}{\psi_{\overline{fb}}(k,t)}$$
$$\cdot (0.5 \cdot \alpha(k+1,t) + \beta(k+1,t)) \cdot 0.5, \qquad (6.58)$$

$$q_{In,b}(k+1,t) = 2 \cdot \pi_n(k,t)\frac{\pi_b(k,t)}{\psi_{\overline{fb}}(k,t)} \cdot 2\alpha(k+1,t) \cdot 0.5, \qquad (6.59)$$

$$q_{In,bb}(k+1,t) = 2 \cdot \pi_n(k,t)\frac{\pi_{bb}(k,t)}{\psi_{\overline{fb}}(k,t)} \cdot 2\beta(k+1,t) \cdot 0.5, \qquad (6.60)$$

$$q_{Inb,nb}(k+1,t) = \pi_{nb}(k,t)\frac{\pi_{nb}(k,t)}{\psi_{\overline{fb}}(k,t)} \cdot 1.5 \cdot \beta(k+1,t) \cdot 0.5, \qquad (6.61)$$

$$q_{Inb,b}(k+1,t) = 2 \cdot \pi_{nb}(k,t)\frac{\pi_b(k,t)}{\psi_{\overline{fb}}(k,t)}$$
$$\cdot (\alpha(k+1,t) + \beta(k+1,t)) \cdot 0.5, \qquad (6.62)$$

$$q_{Inb,bb}(k+1,t) = 2 \cdot \pi_{nb}(k,t)\frac{\pi_{bb}(k,t)}{\psi_{\overline{fb}}(k,t)} \cdot 2\beta(k+1,t) \cdot 0.5, \qquad (6.63)$$

$$q_{Ib,b}(k+1,t) = \pi_b(k,t)\frac{\pi_b(k,t)}{\psi_{\overline{fb}}(k,t)} \cdot 2\alpha(k+1,t) \cdot 0.5, \qquad (6.64)$$

$$q_{Ib,bb}(k+1,t) = 2 \cdot \pi_b(k,t)\frac{\pi_{bb}(k,t)}{\psi_{\overline{fb}}(k,t)} \cdot 2\beta(k+1,t) \cdot 0.5, \qquad (6.65)$$

$$q_{Ibb,bb}(k+1,t) = \pi_{bb}(k,t)\frac{\pi_{bb}(k,t)}{\psi_{\overline{fb}}(k,t)} \cdot 2\beta(k+1,t) \cdot 0.5, \qquad (6.66)$$

$$q_{Ifb,fb}(k+1,t) = \pi_{fb}(k,t)\frac{\pi_{fb}(k,t)}{\psi_{fb}(k,t)} \cdot 2 \cdot 0.5, \tag{6.67}$$

$$q_{Ifb,nbfb}(k+1,t) = 2 \cdot \pi_{fb}(k,t)\frac{\pi_{nbfb}(k,t)}{\psi_{fb}(k,t)}$$
$$\cdot(1 + \beta(k+1,t)) \cdot 0.5, \text{ and} \tag{6.68}$$

$$q_{Inbfb,nbfb}(k+1,t) = \pi_{nbfb}\frac{\pi_{nbfb}}{\psi_{fb}(k,t)} \cdot 2\beta(k+1,t) \cdot 0.5. \tag{6.69}$$

With Eqs. (6.53) to (6.69), the probability $q(k+1,t)$ that a packet is ready to be sent to a destination buffer at stage $k+1$ ($0 \le k \le n-1$), independently of whether the destination buffer is available, can be determined:

$$q(k+1,t) = \frac{q_I(k+1,t)}{\alpha(k+1,t)}. \tag{6.70}$$

Some constraints of the previous equations can also be given. Because the packets will immediately be removed after the network clock cycle in which they arrive at the network outputs ends, we get

$$\alpha(n,t) = 1 \quad \text{and} \tag{6.71}$$
$$\beta(n,t) = 1. \tag{6.72}$$

The probability that a packet arrives at the available buffer of the first network stage is given by the offered load *load* of the network:

$$q(0,t) = load. \tag{6.73}$$

An offered load of 1 accomplishes an immediate pushing of packets into the network if a buffer at the first stage is available.

### 6.3.2 Multiple State Spaces

In addition to the state of the packet in the first position of an SE input buffer, a second state space of the buffer is needed: the buffer queue length. The queue length shows whether a buffer is full, empty, or in a state in between, which is important for determining the probability of packet transmission to a buffer. On the other hand, the first buffer position shows between which packets a conflict occurs, and is thus important for determining the probability of packet transmission to the SE output.

Combining both state spaces into a single state space would lead to a large state space for each buffer. Therefore, the state space of the entire network would become too large to be handled. The beginning of Sect. 6.3 gave an example where ten packet types and a buffer of the size of ten packets would lead to $10^{10}$ states for each SE input. The more concrete example above deals with seven packet types, including the state "no packet" as state 0. But all seven states apply only to the first buffer position. All other positions can only be in state $n$, $b$, or 0 due to the definitions. Nevertheless, if a buffer size of ten packets is used, a large state space still results for an SE input buffer. Considering that an empty buffer position can only be followed by other empty positions, $1 + 6 \cdot (1 + 2 \cdot (1 + 2 \cdot (\ldots))) = 6{,}133$ states describe an SE input buffer. Thus, an $n$-stage MIN would result in $6{,}133^n$ states.

If only the packet type of the first buffer position is considered (because it is the only one relevant) and the type of the remaining buffer positions (but not the queue length) is neglected, the state space is dramatically reduced to $6 \cdot 10 + 1 = 61$ states (six packet types with ten queue lengths each and an empty queue). Then, $61^n$ states describe an $n$-stage network.

Establishing multiple state spaces as presented in Sect. 4.2.2 leads to further state reduction. If the state of the first buffer position and the buffer queue length are separated, two independent state spaces emerge. An SE input buffer is described by $7 + 11 = 18$ states, including the empty queue added to the ten distinguishable queue lengths. A $n$-stage network is described by $18^n$ states.

Of course, the two state spaces are not completely independent, as Eq. (6.51) shows. But keeping the state space of each SE separate from the other ones and combining them during calculation further reduces the number of states to be stored to $18 \cdot n$. The additional calculation time spent for combining states is low, due to only 18 states per network stage.

**State Probabilities of Buffer Queue Lengths**

In the following, the second state space of the MIN model is determined. The probability that the buffer at stage $k$ contains a packet queue of length $m$ ($1 \le m \le m_{max}(k)$, where $m_{max}(k)$ is the buffer length at stage $k$) is called $\pi_m(k, t)$. For instance, $\pi_0(k, t)$ is the probability of a buffer queue of length 0, i.e., that the buffer is empty.

For $2 \le m \le m_{max}(k) - 1$, $m$ packets result in the buffer in three cases. First, if there were $m - 1$ packets before, and no packet left, and a packet was ready to be sent into this buffer. Second, if there were $m$ packets before, and no packet left, and no packet was ready to be sent into this buffer; or if there were $m$ packets before, and a packet left, and another packet was ready to be sent into the buffer. Third, if there were $m + 1$ packets before and a packet left and no packet was ready to be sent into the buffer. Therefore, the probability $\pi_m(k, t)$ results in

$$
\begin{aligned}
\pi_m(k, t) = {} & \pi_{m-1}(k, t-1) \cdot p_{nosend}(k, t-1) \cdot q(k, t-1) \\
& + \pi_m(k, t-1) \cdot (p_{nosend}(k, t-1) \cdot (1 - q(k, t-1)) \\
& \quad + p_{send}(k, t-1) \cdot q(k, t-1)) \\
& + \pi_{m+1}(k, t-1) \cdot p_{send}(k, t-1) \cdot (1 - q(k, t-1)) \\
& \text{if } m_{max}(k) > 1 \ \wedge \ 2 \le m \le m_{max}(k) - 1.
\end{aligned} \tag{6.74}
$$

For $m = 0$, $m = 1$, and $m = m_{max}(k)$, the following particular constraints (see [203]) must be considered:

$$\pi_0(k, t) = \pi_0(k, t - 1) \cdot (1 - q(k, t - 1))$$
$$+ \pi_1(k, t - 1) \cdot p_{send}(k, t - 1) \cdot (1 - q(k, t - 1)), \quad (6.75)$$

$$\pi_1(k, t) = \pi_0(k, t - 1) \cdot q(k, t - 1)$$
$$+ \pi_1(k, t - 1) \cdot (p_{nosend}(k, t - 1) \cdot (1 - q(k, t - 1))$$
$$+ p_{send}(k, t - 1) \cdot q(k, t - 1))$$
$$+ \pi_2(k, t - 1) \cdot p_{send}(k, t - 1) \cdot (1 - q(k, t - 1))$$
$$\text{if } m_{max}(k) > 1 \wedge m = 1, \quad (6.76)$$

$$\pi^{(1)}_{m_{max}(k)}(k, t) = \pi_0(k, t - 1) \cdot q(k, t - 1)$$
$$+ \pi_1(k, t - 1) \cdot (p_{nosend}(k, t - 1)$$
$$+ p_{send}(k, t - 1) \cdot q(k, t - 1))$$
$$\text{if } m_{max}(k) = 1, \text{ and} \quad (6.77)$$

$$\pi^{(2)}_{m_{max}(k)}(k, t) = \pi_{m_{max}(k)-1}(k, t - 1) \cdot p_{nosend}(k, t - 1) \cdot q(k, t - 1)$$
$$+ \pi_{m_{max}(k)}(k, t - 1) \cdot (p_{nosend}(k, t - 1)$$
$$+ p_{send}(k, t - 1) \cdot q(k, t - 1))$$
$$\text{if } m_{max}(k) > 1 \wedge m = m_{max}(k). \quad (6.78)$$

The probability $p_{send}$ that a packet leaves a buffer was calculated before in Eq. (6.49) and is given by

$$p_{send}(k, t) = \beta(k, t), \quad (6.79)$$

hence, the probability that it does not leave is

$$p_{nosend}(k, t) = 1 - p_{send}(k, t). \quad (6.80)$$

## State Probabilities of the First Buffer Position

The first state space, which was introduced at the beginning of this section and includes the packet types, is now determined.

The probability $\pi_n(k, t)$ that the first buffer position at stage $k$ ($0 \leq k \leq n - 1$) contains a normal packet (state $n$) is calculated first. The buffer will reach this state in the next clock cycle in five cases.

First, if there was a normal packet before that did not leave and the destination buffer is not full (otherwise, the state changes to $nb$), for example, because another packet wins an existing conflict. The probability that the destination buffer is not full under the condition that there is at least one packet in the destination buffer (e.g., the winning packet) is given by $p_{notfull}(k+1, t)$.

Second, if there was a broadcast packet before from which a copy left and the destination buffer of the remaining packet is not full.

Third, if there were two or more packets in the buffer (probability $1 - \pi_1(k, t-1) - \pi_0(k, t-1)$), the first one left, and the second one, which now becomes the first one, is a normal packet (probability $\omega_1(k)$), and the destination buffer, which could be empty in this case, is not full. The probability $\omega_1(k)$ that a new packet at stage $k$ is destined to only a single SE output depends on the network traffic pattern. The calculation of $\omega_i(k)$, which gives the probability that $i$ SE outputs are destination of a packet at the SE input, is derived in Sect. 6.1.3.

Fourth, if the buffer was empty and a new packet, which is a normal packet, was sent into the buffer and the destination buffer is not full.

Fifth, if the buffer became empty and a new packet, which is a normal packet, was sent into the buffer and the destination buffer is not full.

$$
\begin{aligned}
\pi_n(k, t) = {} & \pi_n(k, t-1) \cdot (1 - r_n(k, t-1)) \cdot p_{notfull}(k+1, t) \\
& + \pi_b(k, t-1) \cdot r_{pb}(k, t-1) \cdot p_{notfull}(k+1, t) \\
& + (1 - \pi_1(k, t-1) - \pi_0(k, t-1)) \cdot p_{send}(k, t-1) \\
& \quad \cdot \omega_1(k) \cdot (1 - \pi_{m_{max}(k+1)}(k+1, t)) \\
& + \pi_0(k, t-1) \cdot q(k, t-1) \\
& \quad \cdot (\omega_1(k) + \omega_2(k)) \cdot (1 - \pi_{m_{max}(k+1)}(k+1, t)) \\
& + \pi_1(k, t-1) \cdot p_{send}(k, t-1) \cdot q(k, t-1) \\
& \quad \cdot \omega_1(k) \cdot (1 - \pi_{m_{max}(k+1)}(k+1, t)).
\end{aligned}
\tag{6.81}
$$

All other state probabilities of the first buffer position are derived similarly to $\pi_n(k, t)$:

$$
\begin{aligned}
\pi_b(k, t) = {} & \pi_b(k, t-1) \\
& \cdot (1 - r_b(k, t-1) - r_{pb}(k, t-1) \\
& \qquad - r_{pbfb}(k, t-1) - r_{pbfbb}(k, t-1)) \\
& \cdot (1 - p_{full}^2(k+1, t)) \\
& + (1 - \pi_1(k, t-1) - \pi_0(k, t-1)) \cdot p_{send}(k, t-1) \cdot \omega_2(k) \\
& \quad \cdot (1 - p_{full}(k+1, t) \pi_{m_{max}(k+1)}(k+1, t)) \\
& + \pi_0(k, t-1) \cdot q(k, t-1) \cdot \omega_B(k) \cdot (1 - \pi_{m_{max}(k+1)}^2(k+1, t)) \\
& + \pi_1(k, t-1) \cdot p_{send}(k, t-1) \cdot q(k, t-1) \cdot \omega_2(k) \\
& \quad \cdot (1 - p_{full}(k+1, t) \pi_{m_{max}(k+1)}(k+1, t)),
\end{aligned}
\tag{6.82}
$$

$$\pi_{nb}(k,t) = \pi_n(k,t-1) \cdot (1 - r_n(k,t-1)) \cdot p_{full}(k+1,t)$$
$$+\pi_b(k,t-1) \cdot r_{pb}(k,t-1) \cdot p_{full}(k+1,t)$$
$$+\pi_{bb}(k,t-1) \cdot r_{pbb}(k,t-1)$$
$$+\pi_{nb}(k,t-1) \cdot (1 - r_{nb}(k,t-1))$$
$$+\pi_{nbfb}(k,t-1) \cdot (1 - r_{nbfb}(k,t-1))$$
$$\cdot \frac{\pi_{fb}(k,t-1)r_{fb}(k,t-1) + \pi_{nbfb}(k,t-1)r_{nbfb}(k,t-1)}{\psi_{fb}(k,t-1)}$$
$$+(1 - \pi_1(k,t-1) - \pi_0(k,t-1)) \cdot p_{send}(k,t-1)$$
$$\cdot \omega_1(k) \cdot \pi_{m_{max}(k+1)}(k+1,t)$$
$$+\pi_0(k,t-1) \cdot q(k,t-1) \cdot \omega_1(k) \cdot \pi_{m_{max}(k+1)}(k+1,t)$$
$$+\pi_1(k,t-1) \cdot p_{send}(k,t-1) \cdot q(k,t-1)$$
$$\cdot \omega_1(k) \cdot \pi_{m_{max}(k+1)}(k+1,t), \tag{6.83}$$

$$\pi_{bb}(k,t) = \pi_b(k,t-1) \cdot (1 - r_b(k,t-1) - r_{pb}(k,t-1)$$
$$-r_{pbfb}(k,t-1) - r_{pbfbb}(k,t-1)) \cdot p_{full}^2(k+1,t)$$
$$+\pi_{bb}(k,t-1)$$
$$\cdot (1 - r_{bb}(k,t-1) - r_{pbb}(k,t-1) - r_{pbbfb}(k,t-1))$$
$$+(1 - \pi_1(k,t-1) - \pi_0(k,t-1)) \cdot p_{send}(k,t-1) \cdot \omega_2(k)$$
$$\cdot p_{full}(k+1,t)\pi_{m_{max}(k+1)}(k+1,t)$$
$$+\pi_0(k,t-1) \cdot q(k,t-1) \cdot \omega_2(k) \cdot \pi_{m_{max}(k+1)}^2(k+1,t)$$
$$+\pi_1(k,t-1) \cdot p_{send}(k,t-1) \cdot q(k,t-1) \cdot \omega_2(k)$$
$$\cdot p_{full}(k+1,t)\pi_{m_{max}(k+1)}(k+1,t), \tag{6.84}$$

$$\pi_{fb}(k,t) = \pi_b(k,t-1) \cdot r_{pbfb}(k,t-1) \cdot p_{notfull}(k+1,t), \text{ and} \tag{6.85}$$

$$\pi_{nbfb}(k,t) = \pi_b(k,t-1) \cdot r_{pbfbb}(k,t-1)$$
$$+\pi_b(k,t-1) \cdot r_{pbfb}(k,t-1) \cdot p_{full}(k+1,t)$$
$$+\pi_{bb}(k,t-1) \cdot r_{pbbfb}(k,t-1)$$
$$+\pi_{nbfb}(k,t-1) \cdot (1 - r_{nbfb}(k,t-1))$$
$$\cdot \big(\pi_{fb}(k,t-1)(1 - r_{fb}(k,t-1))$$
$$+\pi_{nbfb}(k,t-1)(1 - r_{nbfb}(k,t-1))\big) \frac{1}{\psi_{fb}(k,t-1)}. \tag{6.86}$$

The probability $p_{full}(k,t)$ that the buffer is full is defined for the case where it is known that the buffer is not empty, but it is not known whether it is full or not. This probability is $(1 \le k \le n-1)$

$$p_{full}(k,t) = \frac{\pi_{m_{max}(k)}(k,t)}{1 - \pi_0(k,t)}. \tag{6.87}$$

Then, we get

$$p_{notfull}(k, t) = 1 - p_{full}(k, t). \tag{6.88}$$

Because the packets will immediately be removed after arrival from the outputs of the network, the constraints

$$\alpha(n, t) = 1, \tag{6.89}$$
$$\beta(n, t) = 1, \tag{6.90}$$
$$p_{full}(n, t) = 0, \text{ and} \tag{6.91}$$
$$p_{notfull}(n, t) = 1 \tag{6.92}$$

are given.

### 6.3.3 Fixed Point Iteration

Due to the large number of equations, particularly if the number of network stages increases, a symbolic solution of the system of equations is not feasible. But it can numerically be solved by fixed point iteration.

**Algorithm**

The equations can be solved iteratively, starting with the initialization ($0 \leq k \leq n - 1$):

$$\pi_0(k, 0) = 1, \tag{6.93}$$
$$\pi_n(k, 0) = \pi_b(k, 0) = \pi_{nb}(k, 0) = \pi_{bb}(k, 0)$$
$$= \pi_{fb}(k, 0) = \pi_{nbfb}(k, 0) = 0, \tag{6.94}$$
$$\pi_m(k, 0) = 0 \quad \text{for} \quad 1 \leq m \leq m_{max}(k), \tag{6.95}$$
$$\alpha(k, 0) = \beta(k, 0) = 1, \text{ and} \tag{6.96}$$
$$\psi_{fb}(k, 0) = \psi_{\overline{fb}}(k, 0) = 1. \tag{6.97}$$

First, the traffic distribution in the network ($\omega_1(k)$ and $\omega_2(k)$) is calculated. Then, within the iterative loop, the probabilities $r(k, t)$, $\alpha(k, t)$, and $\beta(k, t)$ (starting with the last network stage) are determined first, followed by all $q_I(k, t)$ and $q(k, t)$, then all $\pi_m(k, t)$, followed by all $p_{full}(k, t)$ and the states of the first buffer position, and finally, all $\psi_{fb}(k, t)$. The algorithm is given below:

get network size (number of stages $n$), buffer sizes $m_{max}$,
    offered load *load*, and traffic pattern $a$;
for $k = 0$ to $n - 1$ do
    calculate $\omega_1(k)$ and $\omega_2(k)$;
end do;
for $k = 0$ to $n - 1$ do
    initialize $\pi_0(k,0)$, $\pi_n(k,0)$, $\pi_{nb}(k,0)$, $\pi_b(k,0)$, $\pi_{bb}(k,0)$,
        $\pi_{fb}(k,0)$, and $\pi_{nbfb}(k,0)$;
    for $m = 1$ to $m_{max}(k)$ do
        initialize $\pi_m(k,0)$;
    end do;
    initialize $\alpha(k,0)$ and $\beta(k,0)$;
    initialize $\psi_{fb}(k,0)$ and $\psi_{\overline{fb}}(k,0)$;
end do;
repeat
    for $k = n - 1$ to 0 step $-1$ do
        calculate all $r(k,t)$;
        calculate $\beta(k,t)$ and $\alpha(k,t)$;
    end do;
    for $k = 1$ to $n$ do
        calculate $q_I(k,t)$ and $q(k,t)$;
    end do;
    for $k = 0$ to $n - 1$ do
        calculate $p_{send}(k,t)$;
        for $m = 0$ to $m_{max}(k)$ do
            calculate $\pi_m(k,t)$;
        end do;
        calculate $p_{full}(k,t)$;
    end do;
    for $k = 0$ to $n - 1$ do
        calculate $\pi_0(k,t)$, $\pi_n(k,t)$, $\pi_{nb}(k,t)$, $\pi_b(k,t)$,
            $\pi_{bb}(k,t)$, $\pi_{fb}(k,t)$, and $\pi_{nbfb}(k,t)$;
    end do;
until steady state has been reached;
calculate the measures;
print the measures;

## Measures

After the steady state has been reached, the normalized throughput at the
network input (called $S_i$) and at the output (called $S_o$) can be determined.
The normalized throughput at the input (output) of the network is the mean
number (expected value) of packets that pass an input (output) of the network
in a network clock cycle.

A new packet is able to enter the network if buffer space is available in the first network stage (probability $1 - \pi_{m_{max}(0)}(0)$) or if the buffer is full and a packet is sent out (probability $\pi_{m_{max}(0)}(0) \cdot p_{send}(0)$). Then, new packets enter the network with the offered load $load$:

$$S_i = (1 - \pi_{m_{max}(0)}(0)) \cdot load$$
$$+ \pi_{m_{max}(0)}(0) \cdot p_{send}(0) \cdot load. \tag{6.98}$$

The normalized throughput at the network output is equal to the probability that a packet is able to be sent to the network output, because the outputs are always available (packets are immediately removed from the outputs):

$$S_o = q(n). \tag{6.99}$$

The normalized delay time at a stage is the average time a packet stays at the stage, and is normalized to the network clock cycle.

The normalized delay time $d(k)$ at each stage $k$ can be calculated using Little's Law. The number of packets in the buffer at stage $k$ that will pass an SE output results from the number in the first buffer position and the number in the remaining buffer positions. Broadcast and blocked broadcast packets generate two packets that will pass an output. Therefore, the mean number in the first buffer position is $\pi_n(k) + \pi_{nb}(k) + \pi_{fb}(k) + \pi_{nbfb}(k) + 2(\pi_b(k) + \pi_{bb}(k))$. Additionally, the mean number in the remaining buffer positions is the probability that there are $m$ packets in the buffer ($2 \leq m \leq m_{max}(k)$) multiplied by $m$ (but one, because the packet in the first buffer position, has already been considered). With probability $w_2(k)$, these packets are broadcast packets, and generate a second packet at the SE outputs. The normalized delay time at stage $k$ is

$$d(k) = \frac{1}{q_I(k+1)} \cdot (\pi_n(k) + \pi_{nb}(k) + \pi_{fb}(k) + \pi_{nbfb}(k) + 2(\pi_b(k) + \pi_{bb}(k))$$
$$+ (1 + \omega_2(k)) \cdot \sum_{m=2}^{m_{max}(k)} ((m-1) \cdot \pi_m(k))). \tag{6.100}$$

The normalized delay time $d_{tot}$ in the network is the sum of all stage delay times:

$$d_{tot} = \sum_{k=0}^{n-1} d(k). \tag{6.101}$$

The mean buffer queue length at a stage is the mean number of packets in the buffer at this stage.

The mean queue length $\bar{m}(k)$ of the buffers at each stage $k$ is given by the probability that there are $m$ packets in the buffer multiplied by $m$ for each possible queue length:

$$\bar{m}(k) = \sum_{m=1}^{m_{max}(k)} (m \cdot \pi_m(k)). \tag{6.102}$$

The mean queue length is a by-product of our algorithm but an important measure in buffer size design.

Some MIN performance results of this mathematical model are presented in Sect. 6.5. Results are determined in computation time of less than one second. The mathematical model achieves a tremendous acceleration in computation time when compared to simulation models, which consume at least four hours (*MINSimulate*) and up to two weeks (Petri nets).

## 6.4 Mathematical Model: Automatic Model Generation

The previous section showed how large the system of equations can grow even if complexity reduction is applied to the model. This means that establishing the system of equations is very time consuming. If changes are required, for instance, if cut-through switching instead of store-and-forward switching is to be used, the entire system of equations must be revised. Such small modifications only slightly change the system of equations (see [209]). But larger modifications, for instance, if $c \times c$ SEs are used instead of simple $2 \times 2$ SEs, again lead to very time consuming development of the system of equations. An automatic model generation as described in Sect. 4.3 would be very helpful.

This section deals with automatic model generation for multistage interconnection networks with the banyan property consisting of $c \times c$ switching elements. $c$ is arbitrary, but must be constant for the entire network. $c \times c$ SEs are chosen to demonstrate how powerful automatic model generation is. The previous section only dealt with $2 \times 2$ SEs to allow manually establishing the system of equations. Larger SEs result in an exponentially growing number of equations related to the increasing number of input-output pairs at each SE. Despite this growth, automatic model generation still succeeds in establishing the system of equations.

### 6.4.1 Rule Design

As discussed in Sect. 4.3, the first step in automatic model generation is to classify groups of similar states or state transitions. Such similar states or state transitions lead to similar equations. The presented example can profit from the equations established in Sect. 6.3. Comparing these equations gives four groups:

1. sending probabilities,
2. receiving probabilities,
3. state probabilities: buffer queue lengths, and
4. state probabilities: first buffer positions.

As the second step in automatic model generation, the rules to establish the equations of each group must be developed. Before the rules are introduced, some definitions are given to adapt the assumptions of Sect. 6.3 to $c \times c$ SEs.

The state space of the first buffer position, which copes with $c \times c$ SEs, represents all reachable states of the first buffer position. It is obvious that packets that are directed to different numbers of SE outputs have to be distinguished. Furthermore, earlier results ([213] and Sect. 6.3) show that the information that blocked packets exist (due to an occupied destination buffer) results in non-negligible information about state probabilities in the next network clock cycle. $\pi_{mult,block}(k,t)$ denotes the probability of being in state $mult, block$ representing a buffer at stage $k$ at time $t$ containing a packet in the first buffer position directed to $mult$ SE outputs. This means that forwarding this packet results in sending copies to $mult$ SE outputs. In the previous clock cycle, sending $block$ of the $mult$ copies ($0 \leq block \leq mult \leq c$) failed because of occupied destination buffers (and at least $mult - block$ copies failed because of lost conflicts with other packets directed to the same destination) if the packet did not arrive newly at the current buffer. For a newly arrived packet, $block$ of the destination buffers are currently occupied. The probability of an empty buffer is denoted as $\pi_{0,0}(k,t)$.

### Rules for Sending Probabilities

First, a packet in the first buffer position at stage $k$ ($0 \leq k \leq n - 1$) at time $t$ is observed that is directed to $mult$ SE outputs, and sending $block$ of the $mult$ copies failed because of occupied destination buffers in the previous clock cycle. $r_{mult,block,nbl,fbl}(k,t)$ denotes the probability that $mult - nbl$ copies of such a packet are sent to the succeeding stage while $nbl$ copies are blocked and stay in the buffer. Those $nbl$ copies include $fbl$ copies that are blocked because of occupied (full) destination buffers. The remaining $nbl - fbl$ copies are blocked because of lost conflicts with other SE inputs for the desired outputs. Figure 6.16 gives an example. The state of the last and the state of the new clock cycle are shown. It is assumed that the packet at the upper buffer is destined to four switch outputs. In the given scenario, two destination buffers of the succeeding stage are full. They block two of the packet copies if no buffer space becomes available. Another destination buffer is also the destination of a second packet. If the second packet wins this conflict, $r_{4,0,3,2}(k,t)$ describes the corresponding sending probability (three blocked copies, two because of full destination buffers).

The sending probability $r_{mult,block,nbl,fbl}(k,t)$ depends on two basic parameters. First, the states of the other SE input buffers influence the number and kind of conflicts. Second, the behavior of the destination buffers influences the availability of the destinations. Again, behaviors $\alpha$, $\bar{\alpha}$, $\beta$, and $\bar{\beta}$ are distinguished for each buffer succeeding one of the SE outputs at stage $k + 1$, as defined in Sect. 6.3.

To automatically generate a sending probability equation, the generator successively examines all combinations of input buffer states and output behaviors. For a specific combination, $P_{blbhv}$ can be derived, which gives the

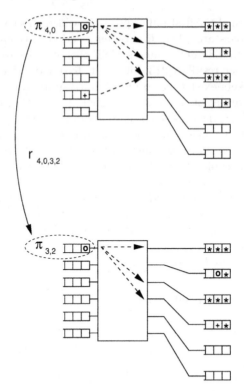

**Fig. 6.16.** Example for sending probability indices

probability that the investigated blocking behavior (*nbl* blocked copies including *fbl* copies that are blocked because of occupied destination buffers) results from this combination.

The sum of the probabilities of each combination multiplied by $P_{blbhv}$ leads to the sending probability $r_{mult,block,nbl,fbl}(k,t)$. Inputs of equal state are accumulated, resulting in the probability $(\pi_{(mult,block)'}(k,t))^{ip_{(mult,block)'}}$ of $ip_{(mult,block)'}$ inputs being in state $(mult,block)'$. To get the probability of a given input combination, the product of all state probabilities represented has to be taken. Figure 6.17 gives an example: let the investigated packet be in conflict with two inputs of state $1,1$ ($ip_{1,1} = 2$) and with three inputs of state $3,2$ ($ip_{3,2} = 3$). All other values of $ip_{(mult,block)'}$ are zero. Therefore, the probability that those five input packets appear in the combination depicted in Fig. 6.17 is $(\pi_{1,1}(k,t))^2 \cdot (\pi_{3,2}(k,t))^3$. Of course, different combinations of the same input packets at inputs 2 to 6 do not change the resulting conflicts. To consider all possible combinations to inputs assignments, the product has to be multiplied by this number of permutations denoted as $Comb_\pi(\{ip_{(mult,block)'}\})$. In the example, $Comb_\pi(\{ip_{(mult,block)'}\}) = \binom{5}{2} = 10$ results. The rules for calcu-

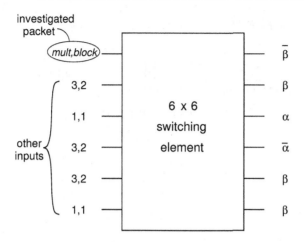

**Fig. 6.17.** Example for input states and output behavior

lating $Comb_\pi(\{ip_{(mult,block)'}\})$ are derived later, after presenting the general rules to set up the equations for $r_{mult,block,nbl,fbl}(k,t)$.

Output behaviors are accumulated in a similar way. For instance, if $op_\alpha$ outputs show behavior $\alpha$, the probability for this is $(\alpha(k+1,t))^{op_\alpha}$. The example (Fig. 6.17) shows $op_\alpha = 1$, $op_{\bar\alpha} = 1$, $op_\beta = 3$, and $op_{\bar\beta} = 1$. The assignment of behaviors to outputs may arbitrarily be permuted. $Comb_{\alpha\beta}(op_\alpha, op_{\bar\alpha}, op_\beta, op_{\bar\beta})$ denotes the number of permutations. Its rules are also derived later, along with the rules for determining the probability $P_{blbhv}$ of obtaining the investigated blocking behavior, given by $nbl$ and $fbl$: it depends on the input states and output behaviors $P_{blbhv} = P_{blbhv}(\{ip_{(mult,block)'}\}, op_\alpha, op_{\bar\alpha}, op_\beta, op_{\bar\beta})$.

To determine a particular $r_{mult,block,nbl,fbl}(k,t)$, all combinations of the input states and output behaviors have to be considered. These rules can be described by Eq. (6.103), and by additionally applying all other equations and algorithms:

$$r_{mult,block,nbl,fbl}(k,t) =$$

$$\sum_{\forall\, combinations} \left( \left( \prod_{\forall\, (mult,block)'} (\pi_{(mult,block)'}(k,t))^{ip_{(mult,block)'}} \right) \right.$$

$$\cdot Comb_\pi(\{ip_{(mult,block)'}\})$$

$$\cdot (\alpha(k+1,t))^{op_\alpha} \cdot (\bar\alpha(k+1,t))^{op_{\bar\alpha}}$$

$$\cdot (\beta(k+1,t))^{op_\beta} \cdot (\bar\beta(k+1,t))^{op_{\bar\beta}}$$

$$\cdot Comb_{\alpha\beta}(op_\alpha, op_{\bar\alpha}, op_\beta, op_{\bar\beta})$$

$$\left. \cdot P_{blbhv}(\{ip_{(mult,block)'}\}, op_\alpha, op_{\bar\alpha}, op_\beta, op_{\bar\beta}) \right),$$

$$(6.103)$$

where

$ip_{(mult,block)'}$ : number of inputs in state $(mult,block)'$ (except examined input)

$Comb_\pi(\{ip_{(mult,block)'}\})$ : number of permutations for given states

$op_\alpha$ : number of outputs with behavior $\alpha$

$op_{\bar\alpha}$ : number of outputs with behavior $\bar\alpha$

$op_\beta$ : number of outputs with behavior $\beta$

$op_{\bar\beta}$ : number of outputs with behavior $\bar\beta$

$Comb_{\alpha\beta}$ : number of permutations for given output behaviors

$P_{blbhv}(\{ip_{(mult,block)'}\}, op_\alpha, op_{\bar\alpha}, op_\beta, op_{\bar\beta})$ :

probability of resulting in the investigated blocking behavior $(nbl, fbl)$.

Because the sending probability of one SE input of $c\times c$ SEs is investigated, there remain $c - 1$ other inputs, so the following constraint holds:

$$\sum_{\forall\, (mult,block)'} ip_{(mult,block)'} = c - 1 \qquad (6.104)$$

The number of permutations $Comb_\pi(\{ip_{(mult,block)'}\})$ according to these inputs for the given input states is

$$Comb_\pi(\{ip_{(mult,block)'}\}) =$$

$$\prod_{\forall\,(mult,block)'} \left( \begin{array}{c} c - 1 - \sum\limits_{(mult,block)''<(mult,block)'} ip_{(mult,block)''} \\ ip_{(mult,block)'} \end{array} \right),$$

$$(6.105)$$

where states $(mult, block)''$ denotes all the states (indices) of the previously already used product terms. As in Eq. (6.104), the output behaviors are constrained by the number of SE outputs:

$$op_\alpha + op_{\bar\alpha} + op_\beta + op_{\bar\beta} = c. \tag{6.106}$$

To calculate $Comb_{\alpha\beta}(op_\alpha, op_{\bar\alpha}, op_\beta, op_{\bar\beta})$, the number of permutations for the given output behaviors, the cases where buffer space is or becomes available (behavior $\alpha$ and $\beta$) and the cases where buffer space does not become available (behavior $\bar\alpha$ and $\bar\beta$) have to be distinguished. $Comb_{\alpha\beta}(op_\alpha, op_{\bar\alpha}, op_\beta, op_{\bar\beta})$ is the number of permutations of those two cases among the outputs for the given behaviors:

$$Comb_{\alpha\beta}(op_\alpha, op_{\bar\alpha}, op_\beta, op_{\bar\beta}) = \binom{c}{op_{\bar\alpha} + op_{\bar\beta}} = \binom{c}{op_\alpha + op_\beta}. \tag{6.107}$$

In the following rules describing $P_{blbhv}(\{ip_{(mult,block)'}\}, op_\alpha, op_{\bar\alpha}, op_\beta, op_{\bar\beta})$, all indices are omitted for notational convenience. To determine $P_{blbhv}$, some assumptions are made without loss of generality (Fig. 6.18):

- The investigated packet is located in the first SE input.

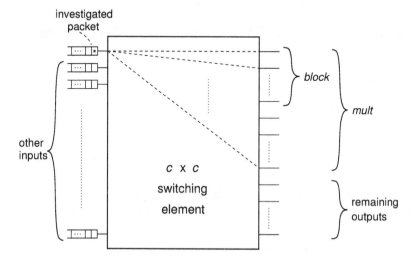

**Fig. 6.18.** Assumptions for an investigated packet of state $mult, block$

- The investigated packet is directed to the first *mult* SE outputs. The first *block* of those outputs are followed by occupied buffers that blocked the copies in the previous clock cycle.

Now, an event tree is constructed describing all conflict situations and output behaviors. The investigation of the event tree is necessary because, in the case of multicasting, the probabilities of resulting conflicts for an output depend on the conflicts arising at the other outputs.

The event tree sequentially considers the first *mult* outputs, starting with the first output. Figures 6.19 to 6.22 explain the event tree construction for one of the first *mult* outputs. Let the considered output be *out*. Now, all other inputs *in* ($2 \leq in \leq c$) are successively examined for whether their copies cause a conflict for output *out* with the investigated packet in input 1. Conflicts that occur because of a copy that was previously blocked due to an occupied destination buffer (edge conf_bl) and conflicts that occur because of other copies (edge conf_nbl) are distinguished. All these cases (and those

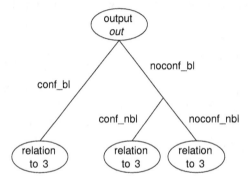

**Fig. 6.19.** Types of conflicts with input 2 (relation to input 3: see Fig. 6.20)

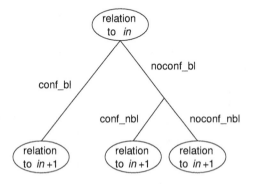

**Fig. 6.20.** Types of conflicts with input *in* ($2 < in < c$)

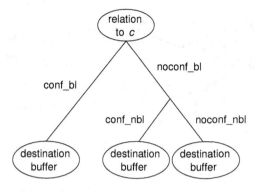

**Fig. 6.21.** Types of conflicts with input $c$

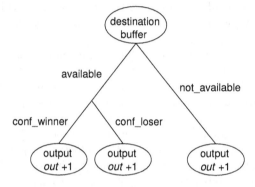

**Fig. 6.22.** Destination availability and conflict resolution

where there is no conflict) are investigated and weighted with their probability of occurring.

Assume that $block_{in}$ copies of the packet staying in input $in$ are destined to outputs followed by occupied destination buffers. Furthermore, assume that $(block_{in})'$ of them are considered to be directed to one of the outputs that have been investigated before output $out$. Edge `conf_bl` occurs if one of the remaining $block_{in,out} = block_{in} - (block_{in})'$ copies is directed to the currently examined output (out of the remaining $c - out + 1$ outputs). The related probability $P_{conf\_bl}$ is

$$P_{conf\_bl} = \frac{block_{in,out}}{c - out + 1}. \tag{6.108}$$

For instance, the packet staying in the second input ($in = 2$) is destined to $block_2 = 2$ occupied destinations (Fig. 6.23). Whether a conflict for the third output ($out = 3$) of a 4×4 SE occurs is investigated. While examining output 1, it might have been assumed that this output was one of the two destinations. Then, examining output 3, there is one of the two copies left

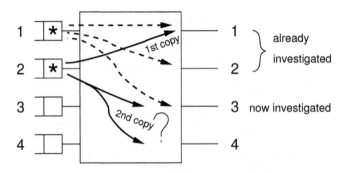

**Fig. 6.23.** Example for conflict scenarios

($block_{2,3} = 1$) that could be destined to output 3 or 4, which have not yet been examined. A conflict with the packet in input 1 for output 3 results with probability $P_{conf\_bl} = 1/2$.

No such previously mentioned conflict (edge `noconf_bl`) occurs with probability $P_{noconf\_bl} = 1 - P_{conf\_bl}$. In this case, there may be a conflict with a copy that was not previously blocked because of an occupied destination buffer (edge `conf_nbl`). Assuming that $mult_{in} - block_{in}$ such copies of input $in$ exist, but $(mult_{in} - block_{in})'$ of them are considered to be directed to one of the outputs that have been investigated before, the remaining $noblock_{in,out} = mult_{in} - block_{in} - (mult_{in} - block_{in})'$ copies may be directed to the current output. The remaining outputs to choose as a destination must be reduced by the number $block_{in,out}$ of outputs that are reserved to be the destination of the blocked copies:

$$P_{conf\_nbl} = \frac{noblock_{in,out}}{c - out + 1 - block_{in,out}}. \tag{6.109}$$

No conflict with such a copy (edge `noconf_nbl`) occurs with probability $P_{noconf\_nbl} = 1 - P_{conf\_nbl}$. No conflict at all occurs with probability $P_{noconf\_bl} \cdot P_{noconf\_nbl}$.

Having considered all inputs and calculated the probability for the chosen path through the event tree, it has to be investigated whether the destination buffer is or becomes available (Fig. 6.22). The probability that the destination buffer is not and does not become available (edge `not_available`) is given by the number $op_{\bar{\alpha}_{out}} + op_{\bar{\beta}_{out}}$ of remaining outputs with behavior $\bar{\alpha}$ or $\bar{\beta}$ that are not allocated to previously examined outputs:

$$P_{not\_available} = \frac{op_{\bar{\alpha}_{out}} + op_{\bar{\beta}_{out}}}{c - out + 1}. \tag{6.110}$$

The destination buffer is or becomes available (edge `available`) with probability $P_{available} = 1 - P_{not\_available}$. Then, the copy of the investigated packet wins the conflict (edge `conf_winner`) with probability

$$P_{conf\_winner} = \frac{1}{inp\_conf + 1}, \qquad (6.111)$$

with $inp\_conf$ denoting the number of other inputs that have a copy directed to the currently examined output for the chosen path in the event tree. The copy of the investigated packet loses the conflict (edge `conf_loser`) with probability $P_{conf\_loser} = 1 - P_{conf\_winner}$.

Now, constructing the event tree is continued, investigating the next output $out + 1$ in the same way. After all $mult$ outputs of the investigated packet in input 1 have been considered, the chosen path through the event tree gives the number of won and lost conflicts for the path and the probability $P_{path}$ for choosing this path.

The remaining $c - mult$ outputs of the $c \times c$ switching element are the destination of the copies injected from input 2 to $c$ that have still not been considered to be directed to one of the previously examined outputs. Copies that have been blocked because of occupied destination buffers result in output behavior $\beta$ or $\bar{\beta}$, and therefore influence the number of outputs with such behavior. But, because $P_{blbhv}$ is derived for a fixed number $op_\beta + op_{\bar{\beta}}$ of outputs with such behavior, the portion of all combinations resulting in the given number must be calculated: input $in$ ($2 \leq in \leq c$) injects $(block_{in})_{rest}$ copies previously not considered into the switching element that have been blocked because of occupied destination buffers. Then,

$$Comb_{bl_{all}} = \prod_{in=2}^{c} \binom{c - mult}{(block_{in})_{rest}} \qquad (6.112)$$

combinations are available to spread these copies over the remaining outputs. But only $Comb_{bl_{fixed}}$ combinations result in the required number of outputs with the described behavior. $Comb_{bl_{fixed}}$ is given by the number of combinations to choose the remaining $op_{\beta_{rest}} + op_{\bar{\beta}_{rest}}$ outputs with behavior $\beta$ or $\bar{\beta}$ from all remaining outputs. Furthermore, it is given by the number of combinations of the remaining blocked copies $((block_{in})_{rest}$ for input $in$) that exactly address these $op_{\beta_{rest}} + op_{\bar{\beta}_{rest}}$ outputs:

$$Comb_{bl_{fixed}} = \binom{c - mult}{op_{\beta_{rest}} + op_{\bar{\beta}_{rest}}}$$
$$\cdot \text{exact}\Big(op_{\beta_{rest}} + op_{\bar{\beta}_{rest}}, \{(block_{in})_{rest} | 2 \leq in \leq c\}\Big),$$
$$(6.113)$$

with

$$\text{exact}\Big(op, \{(block_{in})_{rest}|2 \le in \le c\}\Big) = \prod_{in=2}^{c}\binom{op}{(block_{in})_{rest}}$$

$$- \sum_{j=1}^{op-1}\left(\text{exact}\Big(op - j, \{(block_{in})_{rest}|2 \le in \le c\}\Big) \cdot \binom{op}{op - j}\right).$$

$$(6.114)$$

In a similar way, the given number $op_{\bar{\alpha}} + op_{\bar{\beta}}$ of current blocking behavior must hold. If there remain $op_{\bar{\alpha}_{rest}} + op_{\bar{\beta}_{rest}}$ outputs with such a behavior in the $c - mult$ outputs still not considered,

$$Comb_{bar_{all}} = \binom{c - mult}{op_{\bar{\alpha}_{rest}} + op_{\bar{\beta}_{rest}}} \qquad (6.115)$$

combinations are possible. But only those combinations $Comb_{bar_{fixed}}$ that result in the required $op_{\bar{\beta}_{rest}}$ blockings of outputs to previously occupied destination buffers and $op_{\bar{\alpha}_{rest}}$ blockings of other outputs are considered.

$$Comb_{bar_{fixed}} = \binom{op_{\beta_{rest}} + op_{\bar{\beta}_{rest}}}{op_{\bar{\beta}_{rest}}} \cdot \binom{op_{\alpha_{rest}} + op_{\bar{\alpha}_{rest}}}{op_{\bar{\alpha}_{rest}}}. \qquad (6.116)$$

Taking all paths in the event tree into account, along with Eqs. (6.112) to (6.116), the probability of resulting in the investigated blocking behavior $P_{blbhv}$ is determined by

$$P_{blbhv} = \sum_{\text{all paths}}\left(P_{path} \cdot \frac{Comb_{bl_{fixed}}}{Comb_{bl_{all}}} \cdot \frac{Comb_{bar_{fixed}}}{Comb_{bar_{all}}}\right). \qquad (6.117)$$

Due to the event tree, the calculation of $P_{blbhv}$ is time consuming. The run time to generate the sending probabilities can be reduced using some rules concerning the existence of combinations in Eq. (6.103). If it is known that a combination does not exist, $P_{blbhv}$ does not have to be calculated, and this addend of $r_{mult,block,nbl,fbl}(k,t)$ is set to zero. A combination does *not* exist

- if the number of outputs with behavior $\beta$ or $\bar{\beta}$ is less than the greatest value $block$ that a packet in the first buffer positions of the inputs has.
  Proof: By definition, all $block$ copies are directed to occupied destination buffers. If there exists one input with $block > op_{\beta} + op_{\bar{\beta}}$, then there must be one of the $block$ copies that is directed to an output with behavior $\alpha$ or $\bar{\alpha}$. This contradicts the definition.
- if the added number of outputs with behavior $\beta$ or $\bar{\beta}$ is greater than the sum of all values $block$ of the packets in the first buffer positions of the inputs.
  Proof: By definition, an output shows behavior $\beta$ or $\bar{\beta}$ if there is a copy of a packet that is blocked because of an occupied destination buffer following the investigated output. If $\sum block < op_{\beta} + op_{\bar{\beta}}$, then there must be one output with behavior $\beta$ or $\bar{\beta}$ to which no such copy is directed. This contradicts the definition.

- if the added number of outputs with behavior $\alpha$ or $\beta$ is less than the difference $mult - nbl$ of the investigated packet.

  Proof: If a packet is directed to $mult$ outputs and $nbl$ copies will be blocked in the current clock cycle, the remaining $mult - nbl$ will be sent. Therefore, at least $mult - nbl$ outputs must show either behavior $\alpha$ or $\beta$ to accept those packets.

- if the number of outputs with behavior $\beta$ is less than the difference $block - nbl$ of the investigated packet.

  Proof: If a packet is directed to $mult$ outputs and $block$ copies of it were blocked because of occupied destination buffers, then there exist at most $mult - block$ outputs among the destinations with behavior $\alpha$. If $nbl$ copies will be blocked in the current clock cycle, the remaining $mult - nbl$ will be sent. As stated, at most $mult - block$ copies can be sent to outputs with behavior $\alpha$. Therefore, at least the remaining $(mult-nbl)-(mult-block) = block - nbl$ must be sent to outputs with behavior $\beta$.

- if the number of outputs with either behavior $\bar\alpha$ or $\bar\beta$ is less than the number $fbl$ of copies that will be blocked because of occupied destination buffers.

  Proof: If $op_{\bar\alpha} + op_{\bar\beta} < fbl$, then there are copies among the $fbl$ that are directed to an output with behavior $\alpha$ or $\beta$. This means that those copies will not be blocked. This contradicts the definition of $fbl$.

Considering the above rules and calculation methods, all required equations to determine the sending probabilities $r_{mult,block,nbl,fbl}(k,t)$ are obtained. Some of those equations can further be improved by using additional information. If $nbl$ copies will be blocked, and $fbl$ out of them because of occupied destination buffers, the remaining $nbl - fbl$ copies will be blocked because of lost conflicts. In the previous clock cycle, there were $block$ copies blocked because of occupied destination buffers and $mult - block$ copies blocked because of lost conflicts, and without any information about destination buffer states. As a result, at most $mult - block$ destination buffers are not occupied. If $nbl - fbl > mult - block$, there will exist $(nbl - fbl) - (mult - block)$ copies that are blocked due to lost conflicts but that are also directed to an occupied destination buffer (in addition to the $fbl$ copies). Even if such an occupied destination buffer will send a packet and buffer space becomes available, the winning packet will fill up this space again. As a result, $r_{mult,block,nbl,fbl}(k,t)$ is added to $r_{mult,block,nbl,nbl-(mult-block)}(k,t)$ and then set to zero.

## Rules for Receiving Probabilities

In the previous subsection, rules to determine the sending probabilities were presented. If the sending probabilities are determined, the probability $q_I(k,t)$ that a packet is received by a buffer at stage $k$ ($1 \leq k \leq n$, with stage $n$ representing the network outputs) at time $t$ can easily be calculated by the following rule.

For an input buffer at stage $k-1$ in state $mult, block$, there are $mult - nbl$ copies sent to the succeeding stage, while $nbl$ copies are blocked and remain in the buffer at stage $k-1$ with probability $r_{mult,block,nbl,fbl}(k,t)$. This leads to an average number of $\frac{mult-nbl}{c}$ packets that are received by one of the buffers at stage $k$ from one input buffer at stage $k-1$. Because all $c$ input buffers of the switching element at stage $k-1$ are connected to the investigated buffer at stage $k$, and those $c$ inputs are independent of each other, on average $\frac{mult-nbl}{c} \cdot c = mult - nbl$ packets are received by this buffer. The sum of all states in terms of their occurrence probabilities and sending probabilities results in the receiving probability. Equation (6.118) incorporates these rules:

$$q_I(k,t) = \sum_{\forall\ mult,block} \left( \pi_{mult,block}(k-1,t) \right.$$

$$\left. \cdot \sum_{\forall\ nbl,fbl} \left( r_{mult,block,nbl,fbl}(k-1,t) \cdot (mult - nbl) \right) \right). \quad (6.118)$$

The probability $q(k,t)$ that a packet is ready to be sent into a buffer at stage $k$ $(1 \leq k \leq n)$ at time $t$, regardless of whether the destination buffer is available, is determined similarly to Eq. (6.70) by

$$q(k,t) = \frac{q_I(k,t)}{\alpha(k,t)}. \quad (6.119)$$

The receiving probability $q(0,t)$ of the first network stage is given by the offered load $load$ to the network

$$q(0,t) = load. \quad (6.120)$$

### Rules for State Probabilities of Buffer Queue Lengths

The probability that the buffer at stage $k$ contains a packet queue of length $m$ $(1 \leq m \leq m_{max}(k)$, where $m_{max}(k)$ is the buffer length at stage $k$) is called $\pi_m(k,t)$, as in Sect. 6.3. There are several actions resulting in a buffer queue length of $m$. With regard to the buffer queue length of the previous clock cycle, a length of $m$ results

- if there were $m-1$ packets before, and no packet left, and a packet was ready to be sent into this buffer:

$$\pi_{m_-}(k,t) = \pi_{m-1}(k,t-1) \cdot (1 - p_{send}(k,t-1)) \cdot q(k,t-1)$$
$$\text{if } m_{max}(k) > 1 \ \wedge \ 2 \leq m \leq m_{max}(k) - 1, \quad (6.121)$$

where $p_{send}(k,t)$ denotes the probability of completely sending the packet out of the first buffer position. The rules for $p_{send}(k,t)$ will be determined later. For $m = 1$, the buffer was empty before, and no packet can leave:

$$\pi_{m_-}(k, t) = \pi_{m-1}(k, t-1) \cdot q(k, t-1)$$
$$\text{if } m_{max}(k) \geq 1 \wedge m = 1. \tag{6.122}$$

For $m = 0$, the buffer queue length $m - 1$ does not exist:

$$\pi_{m_-}(k, t) = 0$$
$$\text{if } m_{max}(k) \geq 1 \wedge m = 0. \tag{6.123}$$

- if there were $m$ packets before, and no packet left, and no packet was ready to be sent into this buffer:

$$\pi_{m_0}(k, t) = \pi_m(k, t-1) \cdot (1 - p_{send}(k, t-1)) \cdot (1 - q(k, t-1))$$
$$\text{if } m_{max}(k) > 1 \wedge 1 \leq m \leq m_{max}(k) - 1. \tag{6.124}$$

For $m = m_{max}(k)$, the buffer is full, and no packet can be received by the buffer:

$$\pi_{m_0}(k, t) = \pi_m(k, t-1) \cdot (1 - p_{send}(k, t-1))$$
$$\text{if } m_{max}(k) \geq 1 \wedge m = m_{max}(k). \tag{6.125}$$

For $m = 0$, the buffer is empty, and no packet can be sent:

$$\pi_{m_0}(k, t) = \pi_m(k, t-1) \cdot (1 - q(k, t-1))$$
$$\text{if } m_{max}(k) \geq 1 \wedge m = 0. \tag{6.126}$$

- if there were $m$ packets before, and a packet left, and another packet was ready to be sent into the buffer:

$$\pi_{m_{00}}(k, t) = \pi_m(k, t-1) \cdot p_{send}(k, t-1) \cdot q(k, t-1)$$
$$\text{if } m_{max}(k) \geq 1 \wedge 1 \leq m \leq m_{max}(k). \tag{6.127}$$

For $m = 0$, the buffer is empty, and no packet can be sent:

$$\pi_{m_{00}}(k, t) = 0$$
$$\text{if } m_{max}(k) \geq 1 \wedge m = 0. \tag{6.128}$$

- if there were $m + 1$ packets before, and a packet left, and no packet was ready to be sent into the buffer:

$$\pi_{m_+}(k, t) = \pi_{m+1}(k, t-1) \cdot p_{send}(k, t-1) \cdot (1 - q(k, t-1))$$
$$\text{if } m \neq m_{max}(k). \tag{6.129}$$

For $m = m_{max}(k)$, the buffer queue length $m + 1$ does not exist:

$$\pi_{m_+}(k, t) = 0$$
$$\text{if } m = m_{max}(k). \tag{6.130}$$

The sum of the probabilities related to the four scenarios mentioned describes the state probability of buffer queue length $m$:

$$\pi_m(k, t) = \pi_{m_-}(k, t) + \pi_{m_0}(k, t) + \pi_{m_{00}}(k, t) + \pi_{m_+}(k, t). \tag{6.131}$$

## Rules for State Probabilities of the First Buffer Positions

In this section, the rules to achieve the state probabilities given by the first position of an SE input buffer are derived. $\pi_{mult,block}(k,t)$ denotes the probability of being in state $mult, block$, representing a buffer at stage $k$ at time $t$ containing a packet in the first buffer position directed to $mult$ SE outputs, as defined at the beginning of Sect. 6.4.1.

State $mult, block$ can be reached in two different cases, for which rules have to be developed.

- First, if the packet in the first buffer position of the previous clock cycle is not completely sent to the next network stage due to conflicts and blockings, and the remaining packet obtains state $mult, block$.
- Second, if a new packet that results in state $mult, block$ enters the first buffer position because the buffer is empty or the former packet in the first buffer position has completely left.

If the packet in the first buffer position of the previous clock cycle caused state $mult_{prev}, block_{prev}$, and was not completely sent to the next network stage due to blockings, state $mult, block$ occurs with the related sending probability $r_{mult_{prev},block_{prev},mult,block}$ if such a transition exists (i.e., is not zero):

$$\pi^I_{mult,block}(k,t) = \sum_{\forall\, mult_{prev},block_{prev}} \pi_{mult_{prev},block_{prev}}(k,t-1)$$

$$\cdot r_{mult_{prev},block_{prev},mult,block}(k,t-1). \quad (6.132)$$

The probabilities $r_{mult_{prev},block_{prev},mult,block}$ have been determined in Sect. 6.4.1.

The rules describing the appearance of a new packet in the first buffer position are derived below. Three different buffer lengths are distinguished: the buffer was empty with probability $\pi_0(k,t)$, there was one packet in the buffer with probability $\pi_1(k,t)$, or there were two or more packets in the buffer with probability $1 - \pi_0(k,t) - \pi_1(k,t)$.

- If there were one or more packets in the buffer, a packet may enter the first buffer position if the former packet completely left this position with probability $p_{send}(k,t)$.
- If there were less than two packets in the buffer, a packet may enter the first buffer position if a packet is ready to be send into this buffer with probability $q(k,t)$.
- The probability that a new packet is directed to $mult$ switching element outputs at stage $k$ is denoted as $\omega_{mult}(k)$. This probability results from the given network traffic [207] (see Sect. 6.1.3).
- $block$ destinations must be followed by a full buffer at stage $k+1$, and $mult - block$ destinations must not be followed by a full buffer. All permutations have to be considered.

Expressing these rules as a formula yields:

$$
\begin{aligned}
\pi^{II}_{mult,block}(k,t) = \Big( & \pi_0(k,t-1) \cdot q(k,t-1) \\
& +\pi_1(k,t-1) \cdot p_{send}(k,t-1) \cdot q(k,t-1) \\
& +(1-\pi_0(k,t-1)-\pi_1(k,t-1)) \cdot p_{send}(k,t-1) \Big) \\
& \cdot \omega_{mult}(k) \cdot \Big( \pi_{m_{max}(k+1)}(k+1,t) \Big)^{block} \\
& \cdot \Big( 1-\pi_{m_{max}(k+1)}(k+1,t) \Big)^{mult-block} \\
& \cdot \binom{mult}{block}.
\end{aligned}
\tag{6.133}
$$

The state probabilities of the first buffer position are given by the sum of the previously derived probabilities:

$$
\pi_{mult,block}(k,t) = \pi^{I}_{mult,block}(k,t) + \pi^{II}_{mult,block}(k,t).
\tag{6.134}
$$

### Rules for Buffer Behavior and Measures

This subsection deals with some additional rules to generate some equations out of the four groups mentioned. Each equation is an independent one that does not belong to any group. Nevertheless, it is convenient to automatically generate them depending on the SE size.

To calculate the probability $\beta(k,t)$ that a packet will completely leave an occupied buffer, and the buffer will therefore become available for other packets, each possible buffer state probability $\pi_{mult,block}(k,t)$ and the probability $r_{mult,block,0,0}(k,t)$ of completely sending the packet out of the first buffer position must be considered. Because such a buffer obviously cannot be empty, the probability $\beta(k,t)$ is normalized to non-empty states:

$$
\begin{aligned}
\beta(k,t) = \; & \frac{1}{1-\pi_{0,0}(k,t)} \\
& \cdot \sum_{\forall\, mult,block \neq 0,0} \pi_{mult,block}(k,t) \cdot r_{mult,block,0,0}(k,t).
\end{aligned}
\tag{6.135}
$$

This probability is identical to probability $p_{send}(k,t)$ that a packet is completely sent out of a non-empty buffer:

$$
p_{send}(k,t) = \beta(k,t).
\tag{6.136}
$$

To determine the probability $\alpha(k,t)$ that a packet will not be blocked because of an occupied buffer, two cases have to be distinguished:

- The buffer is not occupied.
- The buffer is occupied, but one packet is leaving.

Expressed as a formula, the rules result in

$$\alpha(k,t) = (1 - \pi_{m_{max}(k)}(k,t)) + \pi_{m_{max}(k)}(k,t) \cdot \beta(k,t). \qquad (6.137)$$

Because the packets will immediately be removed after arrival from the network outputs, the constraints

$$\alpha(n,t) = 1 \quad \text{and} \qquad\qquad\qquad (6.138)$$
$$\beta(n,t) = 1 \qquad\qquad\qquad\qquad (6.139)$$

hold as in Sect. 6.3.2. Network performance is described by the performance measures, as in Sect. 6.3.3: the normalized throughput at the network input and at the output, the delay times, and the buffer queue lengths.

The normalized throughput at the network input is first derived. A new packet is able to enter the network

- if buffer space is available at the first network stage (probability $1 - \pi_{m_{max}(0)}(0)$) or
- if the buffer is full and a packet is sent (probability $\pi_{m_{max}(0)}(0) \cdot p_{send}(0)$).

Then, new packets enter the network with probability $q(0,t)$, which is identical to the offered load *load*:

$$S_i = \left((1 - \pi_{m_{max}(0)}(0)) + \pi_{m_{max}(0)}(0) \cdot p_{send}(0)\right) \cdot q(0,t). \qquad (6.140)$$

The normalized throughput at the network output is equal to the probability that a packet can be sent to the network output because the outputs are always available (packets are immediately removed from the outputs):

$$S_o = q(n). \qquad\qquad\qquad (6.141)$$

The mean delay time at a stage is the average time a packet stays at this stage and is normalized to the network clock cycle. The delay time $d(k)$ at each stage $k$ can be calculated using Little's Law. The number of packets in the buffer at stage $k$ that will pass an SE output results from

- the number in the first buffer position and
- the number in the remaining buffer positions.

A packet that is directed to *mult* outputs generates *mult* copies that will pass an output. Therefore, the mean number in the first buffer position is $\pi_{mult,block}(k) \cdot mult$, considering all states. Additionally, the mean number in the remaining buffer positions results from the probability that there are $m$ packets in the buffer ($2 \leq m \leq m_{max}(k)$) multiplied by $m$ (but the packet in the first buffer position has already been considered). With probability $\omega_{outp}(k)$, these packets are multicast to *outp* outputs, and therefore generate *outp* copies that will pass an output:

$$d(k) = \frac{1}{q_I(k+1)} \cdot \left( \sum_{\forall \, mult,block} \pi_{mult,block}(k,t) \cdot mult \right.$$

$$\left. + \sum_{m=2}^{m_{max}(k)} \left( (m-1) \cdot \pi_m(k) \cdot \sum_{outp=1}^{c} \omega_{outp}(k) \cdot outp \right) \right). \quad (6.142)$$

The mean delay time $d_{tot}$ in the network is the sum of all stage delay times:

$$d_{tot} = \sum_{k=0}^{n-1} d(k). \quad (6.143)$$

The mean buffer queue length at a stage is the average number of packets in the buffer at this stage. The mean queue length of the buffers $\bar{m}(k)$ at each stage $k$ is given by the probability that there are $m$ packets in the buffer multiplied by $m$, for each possible queue length:

$$\bar{m}(k) = \sum_{m=1}^{m_{max}(k)} (m \cdot \pi_m(k)). \quad (6.144)$$

## 6.4.2 Generating and Solving the Equations

The rules above can be used to generate the system of equations to model multistage interconnection networks. To generate the equations and to achieve the results of Sect. 6.5.1, the rules were implemented in C++. The detailed program code is omitted here.

To include the model parameters in the rules, all three different schemes of Sect. 4.3.2 are applied. For instance, the switching technique (store-and-forward) is represented as a hard-coded rule. A model dealing with cut-through switching mainly differs in some additional rules for the sending probabilities and the state probabilities of buffer queue length (see Sect. 6.4.3).

The SE size of the MIN is given as an input parameter during the generation of the equations. The equations are valid only for this SE size.

Network size, network traffic, and buffer size are parameters that are passed during the solution of the equations. The solution can be obtained several times in sequence with changed parameters.

The example presented generates the system of equations as lines of an ordinary programming language, in particular C++. The lines are included in a fixed point iteration algorithm during its compilation. The algorithm works as in Sect. 6.3. Only slight changes in variable names occur. The iteration is initialized with the unloaded network:

$$\pi_0(k,0) = \pi_{0,0}(k,0) = 1, \quad (6.145)$$

$$\pi_m(k,0) = 0 \quad \text{for } 1 \leq m \leq m_{max}(k), \quad (6.146)$$

$$\pi_{mult,block}(k,0) = 0 \quad \text{for } mult, block \neq 0,0, \text{ and} \quad (6.147)$$

$$\alpha(k,0) = \beta(k,0) = 1. \quad (6.148)$$

First, the network parameters, such as network size, buffer sizes, offered load, and traffic pattern, are specified. Then, within the iterative loop, all probabilities $r(k,t)$, $\alpha(k,t)$, and $\beta(k,t)$ (starting with the last network stage) are determined, then all $q_I(k,t)$ and $q(k,t)$, then all $\pi_m(k,t)$, and finally, the states of the first buffer position $\pi_{mult,block}(k,t)$. The algorithm is given below:

```
get network size (number of stages n), buffer sizes m_max(k),
        offered load load, and traffic pattern a;
for k = 0 to n − 1 do
        calculate all ω_outp(k);
end do;
for k = 0 to n − 1 do
        initialize all π_mult,block(k, 0);
        for m = 1 to m_max(k) do
                initialize π_m(k, 0);
        end do;
        initialize α(k, 0) and β(k, 0);
end do;
repeat
        for k = n − 1 to 0 step −1 do
                calculate all r(k, t);
                calculate β(k, t) and α(k, t);
        end do;
        for k = 1 to n do
                calculate q_I(k, t) and q(k, t);
        end do;
        for k = 0 to n − 1 do
                for m = 0 to m_max(k) do
                        calculate π_m(k, t);
                end do;
        end do;
        for k = 0 to n − 1 do
                calculate all π_mult,block(k, t);
        end do;
until steady state has been reached;
calculate the measures;
print the measures;
```

### 6.4.3 Changing the Model

The previous example showed how rules are established and the system of equations is automatically generated. As mentioned in Sect. 4.3, another advantage of establishing equations by rules becomes apparent if the model must be changed. Then, slight changes in the rules for equation generation are usually sufficient to model changes in the system in question.

As an example, the switching technique of the modeled system of this chapter is changed to cut-through switching. Cut-through switching changes the behavior of the MIN as follows: packets that are received by an empty buffer at stage $k$ are immediately (at the same clock cycle) sent to stage $k+1$ if no conflicts with other packets for the related output at stage $k$ occur. This means that packets may pass several network stages per clock cycle. If a conflict occurs, it is solved as between "usual" packets in the store-and-forward switching case.

In the case of store-and-forward switching, packets are always stored in the buffer at the next stage, provided the buffer is not full. This includes the case of an empty buffer. As a result, packets are forwarded at most one network stage per clock cycle.

## Rules for Sending Probabilities

First, the rules for the sending probabilities are determined. Equation (6.103) remains almost unchanged. Only the product term, which is given by $\prod_{\forall\,(mult,block)'}\left(\pi_{(mult,block)'}(k,t)\right)^{ip_{(mult,block)'}}$, has to be adapted. It represents a particular combination of the states of the SE inputs for which an investigated packet wins a potential conflict and is sent. This term is replaced by a term referred to as $InpComb_{ct}$ in the following.

Besides the combinations of Eq. (6.103), which are still valid if no cut-through occurs, additional terms must be considered in the case of a cut-through: an empty buffer may receive a packet that will try to "cut through" to the succeeding stage and that may give rise to a conflict with the packet at the investigated input. This is different from store-and-forward switching, where a conflict with an empty buffer never occurs.

If there are any states $(0,0)'$ of the other inputs (i.e., empty buffers; this means $ip_{(0,0)'} > 0$) and cut-through switching is performed, each of the states may remain in an empty state if no packet is received (probability $1 - q(k,t)$).

If a packet is received (probability $q(k,t)$), it is directed to $mult$ outputs $(1 \leq mult \leq c)$ of the SE at stage $k$ with probability $\omega_{mult}(k)$. A new packet directed to $mult$ outputs has behavior similar to state $(mult,0)'$ (concerning caused output conflicts) and can be treated equally. Conversely, it means that in the case of cut-through switching, each term $\pi_{(mult,0)'}(k,t)$ of the previously mentioned input combinations can be left unchanged to indicate a buffered packet formerly in state $(mult,0)'$, or can be changed to $\pi_{(0,0)'}(k,t) \cdot q(k,t) \cdot \omega_{mult}(k)$ to indicate a "cut-through packet" recently received.

Without loss of generality, it is assumed that $i_{mult}$ of all $ip_{(mult,0)'}$ terms $\pi_{(mult,0)'}(k,t)$ are changed to $\pi_{(0,0)'}(k,t) \cdot q(k,t) \cdot \omega_{mult}(k)$, and the remaining $ip_{(mult,0)'} - i_{mult}$ of these terms remain unchanged $(0 \leq i_{mult} \leq ip_{(mult,0)'})$. Due to the $\binom{ip_{(mult,0)'}}{i_{mult}}$ different ways of choosing the terms to change, such a state for the given $ip_{(mult,0)'}$ SE inputs will occur with probability $\left(\pi_{(mult,0)'}(k,t)\right)^{ip_{(mult,0)'} - i_{mult}} \cdot \left(\pi_{(0,0)'}(k,t) \cdot q(k,t) \cdot \omega_{mult}(k)\right)^{i_{mult}} \cdot \binom{ip_{(mult,0)'}}{i_{mult}}$.

Because an input combination may consist of several terms $\pi_{(mult,0)'}(k,t)$ with different values of $mult$ ($1 \leq mult \leq c$), the change affects all terms $\pi_{(mult,0)'}(k,t)$, and their occurrence probabilities have to be multiplied.

Therefore, the probability $InpComb_{ct}$ of merged input combinations that result in a particular output conflict (i.e., a given input combination without any "cut through," resulting in that particular output conflict and in the added corresponding combination with cut-through occurrence) is determined by

$$InpComb_{ct}(\{ip_{(mult,block)'}\}) =$$

$$\left( \prod_{\forall\ (mult,block)'|_{block>0 \vee mult=0}} \left(\pi_{(mult,block)'}(k,t)\right)^{ip_{(mult,block)'}} \right)$$

$$\cdot (1 - q(k,t))^{ip_{(0,0)'}}$$

$$\cdot \sum_{\forall\ exch} \left( \prod_{mult=1}^{c} \left( \left(\pi_{(mult,0)'}(k,t)\right)^{ip_{(mult,0)'}-i_{mult}} \right. \right.$$

$$\cdot \left(\pi_{(0,0)'}(k,t) \cdot q(k,t) \cdot \omega_{mult}(k)\right)^{i_{mult}}$$

$$\left. \left. \cdot \binom{ip_{(mult,0)'}}{i_{mult}} \right) \right), \tag{6.149}$$

where $exch$ denotes all exchange combinations due to $i_{mult}$. This equation also includes the case where none of the terms $\pi_{(mult,0)'}(k,t)$ is replaced: all $i_{mult}$ are set to $i_{mult} = 0$.

For example, let us assume an SE size of $5 \times 5$ and the term $\pi_{(3,0)'}(k,t) \cdot \pi_{(5,0)'}(k,t) \cdot \pi_{(1,1)'}(k,t) \cdot \pi_{(0,0)'}(k,t) \cdot (1 - q(k,t))$ describing the state of the four other inputs that may be in conflict with the investigated input if no cut-through occurs. Table 6.3 gives the input buffer state probabilities that result in equal output conflicts. The index $(k,t)$ is omitted.

**Table 6.3.** Equal output conflicts

| no cut-through | $\pi_{(3,0)'} \cdot$ | $\pi_{(5,0)'} \cdot$ | $\pi_{(1,1)'} \cdot$ | $\pi_{(0,0)'} \cdot (1-q)$ |
|---|---|---|---|---|
| cut-through | $\pi_{(0,0)'} \cdot q \cdot \omega_3 \cdot$ | $\pi_{(5,0)'} \cdot$ | $\pi_{(1,1)'} \cdot$ | $\pi_{(0,0)'} \cdot (1-q)$ |
| | $\pi_{(3,0)'} \cdot$ | $\pi_{(0,0)'} \cdot q \cdot \omega_5 \cdot$ | $\pi_{(1,1)'} \cdot$ | $\pi_{(0,0)'} \cdot (1-q)$ |
| | $\pi_{(3,0)'} \cdot$ | $\pi_{(5,0)'} \cdot$ | $\pi_{(1,1)'} \cdot$ | $\pi_{(0,0)'} \cdot (1-q)$ |
| | $\pi_{(0,0)'} \cdot q \cdot \omega_3 \cdot$ | $\pi_{(0,0)'} \cdot q \cdot \omega_5 \cdot$ | $\pi_{(1,1)'} \cdot$ | $\pi_{(0,0)'} \cdot (1-q)$ |

Besides the rules on how to establish $InpComb_{ct}$, nothing else is changed, compared to store-and-forward switching.

## Rules for Receiving Probabilities

The rules to establish the receiving probabilities remain unchanged. But another rule must be introduced to deal with an empty buffer at stage $k - 1$ (probability $\pi_{0,0}(k - 1, t)$) that receives a packet (probability $q(k - 1, t)$) and immediately forwards it to some available outputs. The newly received packet at stage $k - 1$ is destined for $mult$ outputs with probability $\omega_{mult}(k - 1)$. Exactly $mult - nbl$ of them are reached with probability $r_{mult,0,nbl,fbl}(k - 1, t)$.

The old rule and the new one lead to the new probability of receiving packets:

$$q_I(k, t) = \sum_{\forall\ mult, block} \left( \pi_{mult,block}(k - 1, t) \right.$$

$$\left. \cdot \sum_{\forall\ nbl, fbl} \left( r_{mult,block,nbl,fbl}(k - 1, t) \cdot (mult - nbl) \right) \right)$$

$$+ \pi_{0,0}(k - 1, t) \cdot q(k - 1, t)$$

$$\cdot \sum_{mult=1}^{c} \left( \omega_{mult}(k - 1) \right.$$

$$\left. \cdot \sum_{\forall\ nbl, fbl} \left( r_{mult,0,nbl,fbl}(k - 1, t) \cdot (mult - nbl) \right) \right). \quad (6.150)$$

All other rules and constraints related to the receiving probabilities remain unchanged.

## Rules for State Probabilities of Buffer Queue Lengths

In the case of cut-through switching, the rules for $\pi_{m_-}(k, t)$ only change for $m = 1$: if the buffer was empty during the last cycle ($m - 1 = 0$), $m = 1$ packets are buffered if the newly received packet, which is directed to $mult$ outputs with probability $\omega_{mult}(k)$, is not completely "cutting through" (probability $1 - r_{mult,0,0,0}(k, t - 1)$):

$$\pi_{m_-}(k, t) = \begin{cases} \pi_{m-1}(k, t - 1) \cdot (1 - p_{send}(k, t - 1)) \cdot q(k, t - 1) \\ \quad \text{if } m_{max}(k) > 1 \wedge 2 \leq m \leq m_{max}(k) - 1 \\ \pi_{m-1}(k, t - 1) \cdot q(k, t - 1) \\ \quad \cdot \sum_{mult=1}^{c} (\omega_{mult}(k) \cdot (1 - r_{mult,0,0,0}(k, t - 1))) \\ \quad \text{if } m_{max}(k) \geq 1 \wedge m = 1 \\ 0 \quad \text{if } m_{max}(k) \geq 1 \wedge m = 0. \end{cases}$$

$$(6.151)$$

In the same way, $\pi_{m_{00}}(k, t)$ changes if $m = 0$: if the buffer was empty during the last cycle ($m = 0$), it remains empty if the newly received packet is completely "cutting through":

$$\pi_{m00}(k,t) = \begin{cases} \pi_m(k,t-1) \cdot p_{send}(k,t-1) \cdot q(k,t-1) \\ \qquad \text{if } m_{max}(k) \geq 1 \wedge 1 \leq m \leq m_{max}(k) \\ \pi_m(k,t-1) \cdot q(k,t-1) \\ \qquad \cdot \sum_{mult=1}^{c} (\omega_{mult}(k) \cdot r_{mult,0,0,0}(k,t-1)) \\ \qquad \text{if } m_{max}(k) \geq 1 \wedge m = 0. \end{cases}$$

$$(6.152)$$

The rules for generating $\pi_{m0}(k,t)$ and for generating $\pi_{m+}(k,t)$ remain unchanged.

### Rules for State Probabilities of the First Buffer Position

The rules to establish the state probabilities of the first buffer position remain unchanged except for the term $\pi_0(k,t-1) \cdot q(k,t-1)$. This term must be removed from $\pi^{II}_{mult,block}(k,t)$. This case of a newly received packet entering the first buffer position because the buffer was empty must be adapted to cut-through switching: if the packet is directed to $mult_{rcvd}$ outputs (probability $\omega_{mult_{rcvd}}(k)$), the buffer state will migrate to this state at the next clock cycle if no cut-through occurs. But a cut-through may occur to the newly received packet, or to some copies of it. Then, $r_{mult_{rcvd},0,mult,block}(k,t-1)$ gives the probability that $mult$ copies remain in the buffer and $block$ of them are blocked because of occupied destination buffers. A third term to be added to Eq. (6.134) emerges:

$$\pi^{III}_{mult,block}(k,t) = \pi_0(k,t-1) \cdot q(k,t-1)$$
$$\cdot \sum_{mult_{rcvd}=1}^{c} (\omega_{mult_{rcvd}}(k) \cdot r_{mult_{rcvd},0,mult,block}(k,t-1)).$$

$$(6.153)$$

All other rules remain unchanged.

### Rules for Buffer Behavior and Measures

All rules for buffer behavior and measures remain unchanged. But a new measure $P_{cut}(k,t)$ can be introduced. It denotes the probability that a newly received packet is completely and immediately forwarded at stage $k$ at time $t$. Such a cut-through occurs if the buffer at stage $k$ is empty (probability $\pi_0(k,t)$) and the newly received packet, which is destined to $mult$ outputs with probability $\omega_{mult}(k)$, is completely sent (probability $r_{mult,0,0,0}(k,t)$):

$$P_{cut}(k,t) = \pi_0(k,t) \cdot \sum_{mult=1}^{c} (\omega_{mult}(k) \cdot r_{mult,0,0,0}(k,t)). \qquad (6.154)$$

## Generating and Solving the Equations

Due to the rule-based modeling and due to the automatically generated equations, no changes are needed for generating and solving the equations (the changed rules however, must be entered into the generator).

## 6.5 Model Engineering and Performance

Previous sections presented four ways to determine the performance of multi-stage interconnection networks. This section summarizes previous results in a comparison. Furthermore, it gives some examples of investigations that were enabled by these models and that lead to interesting results.

### 6.5.1 Comparison of the Modeling Techniques

As multicasting is considered in the models, many different assumptions about the shape of the distribution in space of the network traffic pattern are possible. The simplest is to assume that all possible combinations of destination addresses are equally distributed for each packet entering the network. This traffic pattern is applied in the following results. It determines the multi-cast destination distribution $a(i)$, which leads to the multicast probabilities $\omega_{mult}(k)$ (see Sect. 6.1.3) required for the Petri net model and the mathematical models. Simulation with *MINSimulate* directly deals with $a(i)$. $a(i)$ gives the probability that a packet entering the network has $i$ destination addresses ($1 \leq i \leq N$). The above assumption that all possible combinations of destination addresses are equally distributed yields

$$a(i) = \frac{\binom{N}{i}}{\sum_{j=1}^{N} \binom{N}{j}}. \tag{6.155}$$

Figures 6.24 and 6.25 compare the results of all modeling techniques presented in this chapter. They show the normalized throughput of the network outputs and the normalized delay time. The network size is varied. A buffer size of 1 at each network stage and an offered load of 1 are used. An SE size of 2×2 is chosen. Because of the time-intensive simulation runs, the Petri net models are only determined up to a size of 64 network inputs. All simulations (Petri nets and *MINSimulate*) were stopped when the confidence level reached 95% and the precision reached 2%. The iterative Petri net model starts its first step with a confidence level of 80% and precision of 10% as termination criteria (Sect. 6.1.2). After the results of two consecutive iterations differ by less than 5%, it is moved to the second step with a confidence level of 95% and a precision of 2%.

As Fig. 6.24 shows, the throughput results demonstrate a good correspondence between the various techniques. Only the iterative Petri net description

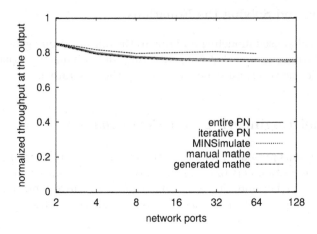

**Fig. 6.24.** Result comparison (normalized throughput)

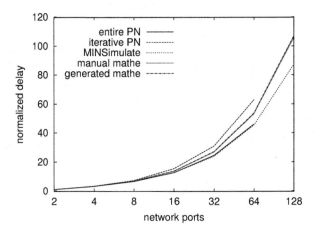

**Fig. 6.25.** Result comparison (normalized delay)

noticeably differs from the other methods. It is caused by the sum up of the inaccuracies of each simulation in the iteration.

The mathematical model that was manually set up (Sect. 6.3) and the automatically generated mathematical model (Sect. 6.4) differ slightly. The models are not completely identical. One of the two main differences is the neglect of some states resulting from conflicts between two broadcast packets in the first buffer positions. These states are called $fb$ and $nbfb$ in the manually set up model. The automatically generated model generally neglects conflictless states resulting from earlier conflicts of multicast packets. Taking such conflicts into account would significantly increase the state space in the case of $c \times c$ switching elements. Many different combinations of output overlapping between all multicast packets in the inputs are feasible. Of course, this state reduction decreases the accuracy of the model. But experiments show that the larger the SE size becomes, the less the state reduction influences accuracy.

The second main difference between the models increases accuracy, because new states have been defined that increase the state space, but only linearly with the SE size, in contrast to approximately exponential growth in the case of conflictless states. The newly introduced states describe how many of the copies of a multicast packet were previously blocked by occupied destinations and how many by lost conflicts. The model that was manually set up only distinguishes between multicast packets blocked by at least one copy directed to an occupied destination and multicast packets not blocked at all by occupied destinations.

Removing the conflictless states decreases accuracy somewhat more than the introduction of new states describing the multicast destination behavior can increase accuracy. There is quite no difference with respect to delay times (Fig. 6.25) between the two models.

Table 6.4 outlines the quality of the results. They are compared to *MINSimulate* as a reference that models the network closest to reality. The quality is determined by the maximum difference between the throughput of the related model and *MINSimulate*.

Table 6.4 also shows the model development times, the computation times (for a 64×64 MIN), and the required memory. The estimated model develop-

**Table 6.4.** Comparison of the modeling methods

| Technique | Development time | Computation time | Quality | Required mem. |
|---|---|---|---|---|
| entire Petri net | 100 person hours | > 2 weeks | 0.2% | 8.6 MByte |
| iterative Petri net | 200 person hours | 20 hours | 4.5% | 5.5 MByte |
| *MINSimulate* | 400 person hours | 4 hours | — | 7.1 MByte |
| manual mathe | 1,500 person hours | < 1 second | 0.2% | 0.9 MByte |
| generated mathe | 400 person hours | < 1 second | 1.3% | 0.9 MByte |

ment times include the time for validation and error detection. There are large differences in the development times. Fast development and model validation characterizes the Petri net descriptions. The main reason for this is the graphical model interface. It allows a proper model survey. The token game supports easy model validation. Because of the decomposition approach, the iterative Petri net description suffers from higher development time. The detection of near-independent subnets to decompose the network is time intensive and requires the intuition of the model designer.

The time spent for the *MINSimulate* program development is higher than that for Petri nets because model validation is more complicated. The validation requires a step-by-step simulation. The given value only represents the time required to establish the basic simulation model that is able to handle the same kinds of MINs as the Petri net models or the mathematical models. Of course, all additional features of *MINSimulate* need additional development time.

The development time of the mathematical model that was manually set up is much higher than in all other methods. The reason is the low-level development based on Markov chains. The state transition probabilities result in complex equations. Development and validation is very time consuming. Automatic model generation significantly reduces the development time.

The computation times (run times) of the methods are as different as the development times, but in reverse order. The mathematical models are the fastest techniques in result achievement. All other models use simulation, which increases the computation time. The *MINSimulate* computation time is lower than the two Petri net computation times because it profits from the program particularly adapted to model multistage interconnection networks. The lower computation time of the iterative Petri net description, compared to the entire Petri net description, results from the smaller model because only one row of the network has to be modeled. Table 6.4 shows the computation times on a 1,200 MHz processor.

### 6.5.2 Model Capabilities

This section briefly presents the capabilities of the models. For detailed studies, it is referred to the related publications.

The models of multistage interconnection networks are developed to investigate various architecture issues and traffic features. For instance, buffer sizes and their influence on network performance are examined in [213, 216, 206]; simulation and mathematical models are used. Different buffer architectures are compared by simulation in [210]. [207, 208, 216, 217] deal with the switching element size and propose a size for optimal performance using mathematical modeling and simulation. Time-dependent traffic is applied to MINs in [204, 205, 215], while [202, 201, 211, 212] show the traffic distribution in space; again both, simulation and mathematical methods are used. In [209], various switching techniques are compared to evaluate their power. The features of MIN architectures with no banyan property are presented in [214, 218, 219] using simulation.

# 7

## Concluding Remarks

In recent years, many parallel and distributed systems have been established. There are several reasons for this development. For instance, decreasing communication costs allow intense use of communication networks. Furthermore, communication bandwidth has been increasing. Thus, large amounts of data can be transferred between the nodes of parallel and distributed systems in reasonable time. Applications are no longer limited to centralized systems.

Powerful management software supports parallel and distributed computing in homogenous systems as well as in highly heterogeneous ones. Particularly, the efforts in ubiquitous computing result in heterogeneous and asymmetric distributed systems. These include wireless networks. Thus, a distributed system may consist of fixed nodes as well of mobile nodes.

This book has dealt with the engineering of network architectures for parallel and distributed systems. The engineering part has been realized by modeling the system under consideration. A performance evaluation of different model parameter sets allows for comprehensive comparison of various architectures in order to find the optimal one concerning the design goal.

A characterization of network architectures for parallel and distributed systems explains features that influence the system behavior, and, therefore, its performance. The network represents one of the most important components. For instance, different switching techniques result in different blocking behaviors, and delay times, particularly, vary. Depending on the applications that run on the system, the network has to deal with various traffic patterns. Traffic patterns are distributed in space and in time. For instance, multicasting shows a distribution in space and has been considered in the examples.

Different kinds of network architecture have also been discussed. Due to their advantages and drawbacks, network performance changes with architecture. Wired and wireless architectures have been distinguished. A separate section has dealt with network-on-chip (NoC) architectures. In wired architectures, multistage interconnection networks are the focus because they reach high performance with low hardware complexity. Investigating multicast traffic has led to a new architecture to support this kind of traffic in an optimal way. The newly described architecture is called multilayer multistage interconnection network. It significantly improves network performance in the case of multicast traffic while only slightly increasing hardware costs.

NoCs, as a special kind of wired architecture, have been characterized by their particular features due to their single chip realization.

To evaluate the performance of parallel and distributed systems, many methods can be applied. This book has concentrated on simulation and mathematical methods. Simulation provides a powerful modeling method with the capability to describe all kinds of systems. But it often suffers from very time-consuming computation times. On the other hand, mathematical methods like Markov chains and Petri nets profit from their short computation time but come with the drawback of limited system description capabilities. Furthermore, model development often becomes a complex task.

Thus, this book has tried to help with model engineering. A guideline on how to choose the right modeling method has been presented. Depending on the modeling method, help with model development has been provided. Model complexity reduction proves to be one of the most important tasks in modeling. A less complex model usually results in shorter computation time, or allows computation for the first time.

Finally, a new concept for automatic model generation significantly accelerates model development. The given strategy derives an automatic generator by considering previously elaborated rules describing system behavior.

Two examples have shown how to apply the above mentioned concepts. A smaller example has dealt with a wireless cellular network. It has examined the handoff procedure of mobile nodes carrying real-time traffic. Due to the moderate model size, a Petri net description sufficiently handles the system in question.

Unfortunately, the second example is more complex. Multistage interconnection networks (MINs) were modeled to optimize their architecture. MINs are of interest due to their use in parallel computers and in switches connecting distributed systems.

It turns out that a MIN is too complex to be simply modeled by Petri nets. The Petri net model is to large to be handled by a computer. Other techniques like simulation and Markov chains have also been applied and compared. The automatic model generation is particularly impressive with its powerful features. For instance, it heavily accelerated model development and system changes could easily be handled.

The automatic model generation will be used in future to develop and investigate reconfigurable parallel and distributed system architectures. Current research examines how reconfigurable networks can optimize network performance.

Bidirectional multistage interconnection networks are of particular interest for reconfigurable parallel systems. They show some locality in communication delay times. Nodes that are located close together (concerning the network inputs) can exchange messages by passing only few stages. In the borderline case, only the first stage is used. But nodes that are connected to network inputs far away from each other must communicate by exchanging messages that pass almost all stages of the MIN.

Reconfiguration raises a couple of questions. For instance, how do we determine the optimal new architecture of the network? Another question concerns dynamic reconfiguration: how do we deal with messages that have not yet finished their path to the destination node at the instant of reconfiguration? This question mainly targets appropriate (e.g., adapting) routing methods. Simple routing algorithms as used in the examples of this book will fail.

Another question asks for the optimal point in time for reconfiguration. Network traffic prediction would be helpful. And because network traffic is strongly related to the applications running on the system, a kind of application prediction must be established.

Important future work will be to include all the issues above into a model of the network to investigate reconfiguration. In contrast to previous models, the network behavior in time becomes very important. Either transient solution techniques of mathematical models or terminating simulations must be applied. Models will no longer refer to a static network structure; dynamic change of the structure must somehow be incorporated into the model description. Adaptive routing techniques and traffic prediction must also be modeled.

At TU Berlin, a new simulator is currently under development to support previous constraints. This simulator is called *CINSim* (component-based interconnection network simulator) [221, 222]. In addition to reconfigurable systems, it also focuses on modeling heterogeneous and asymmetric systems. These kinds of systems cannot be described by *MINSimulate*.

Asymmetric systems require additional effort in routing. Global routing quickly becomes a bottleneck for a large parallel or distributed system. Local routing avoids this, but its modeling raises many more problems than in the case of symmetric systems. The reason is the usually sequential execution of the simulator. This possibly means that events occurring in parallel in a system must somehow be sequentialized. For instance, events interacting and occurring in the stages of a symmetric MIN in parallel can be sequentialized by starting with the switching elements at the last stage and dealing step-by-step with each stage until the first stage with the largest distance to the network outputs is reached. In an asymmetric MIN, a switching element may have different distances to the different network outputs. Therefore, no simple sequentializing method can be applied. If the system in question is additionally heterogeneous, the problem may become even more complex if, for instance, switching elements or routers are clocked with different frequencies due to different hardware constraints.

As previously pointed out, many unsolved problems exist in modeling and evaluating network architectures for parallel and distributed systems. The more complex the architectures of systems become, the more difficult modeling becomes.

# References

1. Abandah, G. A. and Davidson, E. S., Modeling the communication performance of the IBM SP2. In: *Proceedings of the 10th International Parallel Processing Symposium (IPPS'96); Hawaii.* IEEE Computer Society Press, 1996.
2. Abandah, G. A. and Davidson, E. S., Characterizing distributed shared memory performance: A case study of the Convex SPP1000. *IEEE Transactions on Parallel and Distributed Systems,* 9(2):206–216, February 1998.
3. Alderighi, M., Casini, F., D'Angelo, S., Salvi, D., and Sechi, G. R., A fault-tolerant FPGA-based multi-stage interconnection network for space applications. In: *Proceedings of the First IEEE International Workshop on Electronic Design, Test and Applications (DELTA'02),* pp. 302–306, 2002.
4. Allen, D., *Probability, Statistics and Queueing Theory with Computer Science Applications.* Academic Press, New York, 2nd ed., 1990.
5. Atiquzzaman, M. and Akhtar, M. S., Performance of buffered multistage interconnection networks in a nonuniform traffic environment. *Journal of Parallel and Distributed Computing,* 30(1):52–63, October 1995.
6. Awdeh, R. Y. and Mouftah, H. T., Survey of ATM switch architectures. *Computer Networks and ISDN Systems,* 27:1567–1613, 1995.
7. Awdeh, R. Y. and Mouftah, H. T., The expanded delta fast packet switch. In: *Proceedings IEEE SUPERCOMM/ICC'94; New Orleans,* pp. 397–401. IEEE Computer Society, IEEE Computer Society Press, May 1994.
8. Balbo, G., Chiola, G., Franceschinis, G., and Molinar Roet, G., On the efficient construction of the tangible reachability graph of generalized stochastic Petri net models. In: *Proceedings of the 2nd International Workshop on Petri Nets and Performance Models; Madison,* pp. 136–145, 1987.
9. Banks, J., Carson, J. S., Nelson, B. L., and Nicol, D. M., *Discrete-Event System Simulation.* Prentice Hall, 3rd ed., 2000.
10. Bansal, N., Gupta, S., Dutt, N., Nicolau, A., and Gupta, R., Network topology exploration of mesh-based coarse-grain reconfigurable architectures. In: *Proceedings of the Design, Automation and Test in Europe Conference and Exhibition (DATE'04),* pp. 474–479. IEEE, 2004.
11. Becker, J., Dynamically and partially reconfigurable architectures. *it – Information Technology,* 46(4):218–225, 2004.

12. Begain, K., Herold, H., and Bolch, G., Analytical model of cellular mobile networks with adaptive data connections. In: *Proceedings of the European Simulation Symposium 1998 (ESM'98); Manchester*, pp. 787–793. SCS, SCS International, 1998.

13. Beneš, V. E., Optimal rearrangeable multistage connecting networks. *Bell System Technology Journal*, 43:1641–1656, March 1964.

14. Beneš, V. E., *Mathematical Theory of Connecting Networks and Telephone Traffic*, volume 17 of *Mathematics in Science and Engineering*. Academic Press, New York, 1965.

15. Benini, L. and De Micheli, G., Networks on chips: A new SoC paradigm. *IEEE Computer*, 35(1):70–80, 2002.

16. Bertozzi, D. and Benini, L., Xpipes: A network-on-chip architecture for gigascale system-on-chip. *IEEE Circuits and Systems Magazine*, 4(2):18–31, 2004.

17. Bertozzi, D., Jalabert, A., Murali, S., Tamhankar, R., Stergiou, S., Benini, L., and De Micheli, G., NoC synthesis flow for customized domain specific multiprocessor systems-on-chip. *IEEE Transactions on Parallel and Distributed Systems*, 16(2):113–129, February 2005.

18. Bhattacharya, S., Elsesser, G., Tsai, W.-T., and Du, D.-Z., Multicasting in generalized multistage interconnection networks. *Journal of Parallel and Distributed Computing*, 22(7):80–95, July 1994.

19. Bolch, G., Greiner, S., de Meer, H., and Trivedi, K. S., *Queueing Networks and Markov Chains – Modeling and Performance Evaluation with Computer Science Applications*. John Wiley & Sons, New York, 1998.

20. Bolotin, E., Cidon, I., Ginosar, R., and Kolodny, A., Cost considerations in network on chip. *Integration, the VLSI Journal*, 38:19–42, 2004.

21. Boura, Y. M. and Das, C. R., Performance analysis of buffering schemes in wormhole routers. *IEEE Transactions on Computers*, 46(6):687–694, June 1997.

22. Braden et al., Resource reservation protocol - version 1 functional specification. RFC 2205, IETF, 1997.

23. Bratley, P., Fox, B.L., and Schrage, L.E., *A Guide to Simulation*. Springer Verlag, 2nd ed., 1987.

24. Brenner, M., Tutsch, D., and Hommel, G., Measuring transient performance of a multistage interconnection network using Ethernet networking equipment. In: *Proceedings of the International Conference on Communications in Computing 2002 (CIC'02); Las Vegas*, pp. 211–216. CSREA Press, 2002.

25. Breuer, L., *From Markov Jump Processes to Spatial Queues*. Kluwer Academic Publishers, 2003.

26. Burk, W.H., Limitations to parallel processing. In: *Proceedings of the 9th International Phoenix Conference on Computers and Communications*, pp. 86–93. IEEE Press, 1990.

27. Cali, F., Conti, M., and Gregori, E., IEEE 802.11 wireless LAN: Capacity analysis and protocol enhancement. In: *Proceedings of INFOCOM'98; San Francisco*, 1998.

28. Carlsson, P. and Fiedler, M., Multifractal products of stochastic processes. In: *Proceedings of the 15th Nordic Teletraffic Seminar (NTS-15)*, 1999.

29. Castelluccia, C., Extending Mobile IP with adaptive individual paging: A performance analysis. *Mobile Computing and Communications Review*, 5(2):14–26, 2001.

30. Chan, K.-S., Yeung, K. L., and Chan, S., A refined model for performance analysis of buffered banyan networks with and without priority control. In: *Proceedings IEEE Global Telecommunications Conference 97 (GLOBECOM'97); Phoenix*, pp. 1745–1750. IEEE Computer Society, IEEE Computer Society Press, 1997.

31. Chaya, H.S., and Gupta, S., Performance modeling of the asynchronous data transfer methods of IEEE 802.11 MAC protocol. *Wireless Networks*, 3:217–234, 1997.

32. Cheemalavagu, S. and Malek, M., Analysis and simulation of banyan interconnection networks with 2×2, 4×4 and 8×8 switching elements. In: *Proceedings Real-Time Systems Symposium; Los Angeles*, pp. 83–89. IEEE Computer Society, IEEE Computer Society Press, December 1982.

33. Chen, C.-K., and Atiquzzaman, M., An improved model for the performance analysis of multistage switches. In: D. Dowd and E. Gelenbe, editors, *Proceedings of the Third International Workshop on Modeling, Analysis, and Simulation of Computer and Telecommunication Systems; Durham*, pp. 105–109. IEEE Computer Society, IEEE Computer Society Press, January 1995.

34. Cheng, R.C.H., Variance reduction methods. In: *Proceedings of the 1986 Winter Simulation Conference*, pp. 60–68. SCS, 1986.

35. Ching, D., Schaumont, P., and Verbauwhede, I., Integrated modeling and generation of a reconfigurable network-on-chip. In: *Proceedings of the 18th International Parallel and Distributed Processing Symposium (IPDPS 2004)*, pp. 139–145, 2004.

36. Ciardo, G., Blakemore, A., Chimento, P. F., Muppala, J. K., and Trivedi, K. S., Automated generation and analysis of Markov reward models using stochastic reward nets. In: C. Meyer and R. J. Plemmons, editors, *Linear Algebra, Markov Chains, and Queueing Models*, volume 48 of *IMA Volumes in Mathematics and its Applications*, pp. 145–191. Springer Verlag, Berlin, 1993.

37. Clos, C., A study of nonblocking switching network. *Bell System Technology Journal*, 32:406–424, March 1953.

38. Compton, K. and Hauck, S., Configurable computing: A survey of systems and software. *ACM Computing Surveys*, 34(2):177–210, 2002.

39. Corson, S. and Macker, J., Mobil ad-hoc networking (MANET): Routing protocol performance issues and evaluation considerations. RFC 2501, IETF, 1999.

40. Coulouris, G., Dollimore, J., and Kindberg, T., *Distributed Systems: Concepts and Design*. Addison Wesley, 3rd ed., 2001.

41. Courtois, P., Decomposability, instabilities, and saturation in multiprogramming systems. *Communications of the ACM*, 18(7):371–377, July 1975.

42. Courtois, P., *Decomposability: Queueing and Computer System Applications*. Academic Press, New York, 1977.

43. Courtois, P. and Semal, P., Bounds for positive eigenvectors of non-negative matrices and their approximations by decomposition. *Journal of the ACM*, 31(4):804–825, 1984.

44. Cox, D.R. and Miller, H.D., *The Theory of Stochastic Processes*. Chapman & Hall, London, 1968.

45. Culler, D.E., Singh, J.P., and Gupta, A., *Parallel Computer Architecture: A Hardware Software Approach*. Morgan Kaufmann Publishers, 1999.

46. Dally, W. J., and Lacy, S., VLSI architecture: Past, present, and future. In: *Proceedings of the 20th Anniversary Conference on Advanced Research in VLSI*, pp. 232–241, 1999.

47. Dally, W. J. and Towles, B., Route packets, not wires: On-chip interconnection networks. In: *Proceedings of Design Automation Conference (DAC 2001)*, pp. 684–689, 2001.

48. de Rose, C.A.F. and Heiß, H.-U., Dynamic processor allocation in large mesh-connected multicomputers. In: *Proceedings of the EURO-PAR 2001; Manchester; Lecture Notes in Computer Science (LNCS 2150)*. Springer Verlag, 2001.

49. Dias, D. M. and Jump, J. R., Analysis and simulation of buffered delta networks. *IEEE Transactions on Computers*, C–30(4):273–282, April 1981.

50. Ding, J. and Bhuyan, L. N., Finite buffer analysis of multistage interconnection networks. *IEEE Transactions on Computers*, 43(2):243–247, February 1994.

51. Duato, J., Yalamanchili, S., and Ni, L., *Interconnection Networks – An Engineering Approach*. Morgan Kaufmann Publishers, 2003.

52. Tran-Gia et al. Impacts of new services on the architecture and performance of broadband networks. Final Report COST-257, European Union, Universität Würzburg, 2000.

53. Feller, W., *An Introduction to Probability Theory and its Applications*, volume 1. John Wiley & Sons, New York, 3rd ed., 1968.

54. Feller, W., *An Introduction to Probability Theory and its Applications*, volume 2. John Wiley & Sons, New York, 2nd ed., 1971.

55. Feng, T. Y., A survey of interconnection networks. *Computer*, pp. 12–27, December 1981.

56. Fishman, G., Problems in the statistical analysis of simulation experiments: The comparison of means and the length of sample records. *Communications of the ACM*, 10(2):94–99, 1967.

57. Fishman, G. S., *Discrete-Event Simulation*. Springer Verlag, 2001.

58. Fong, S. and Singh, S., Queuing analysis of shared-buffer switches with control scheme under bursty traffic. *Computer Communications*, 21:1681–1692, 1998.

59. Fujii, H., Yasuda, Y., Akashi, H., Inagami, Y., Koga, M., Ishihara, O., Kashiyama, M., Wada, H., and Sumimoto, T., Architecture and performance of the Hitachi SR2201 massively parallel processor system. In: *Proceedings of the 11th International Parallel Processing Symposium (IPPS'97); Genf*, pp. 233–241. IEEE Computer Society Press, April 1997.

60. Fujimoto, R., Parallel discrete event simulation. *Communications of the ACM*, 33:30–60, 1990.

61. Gelenbe, E. and Pujolle, G., *Introduction to Queueing Networks*. John Wiley & Sons, 2nd ed., 1998.

62. German, R., *Analysis of Stochastic Petri Nets with Non-Exponentially Distributed Firing Times*. PhD thesis, Technische Univeristät Berlin, 1994.

63. German, R., *Performance Analysis of Communication Systems*. John Wiley & Sons, 2000.

64. German, R. and Heindl, A., Performance evaluation of IEEE 802.11 wireless LANs with stochastic Petri nets. In: *Proceedings of the 8th Int. Workshop on Petri Nets and Performance Models*, pp. 44–53. IEEE, IEEE Computer Society Press, 1999.

65. German, R., Kelling, C., Zimmermann, A., and Hommel, G., TimeNET — a toolkit for evaluating non–Markovian stochastic Petri nets. *Performance Evaluation*, 24:69–87, 1995.

66. Gharsalli, F., Baghdadi, A., Bonaciu, M., Majauskas, G., Cesario, W., and Jerraya, A. A., An efficient architecture for the implementation of message passing programming model on massive multiprocessor. In: *Proceedings of the 15th IEEE International Workshop on Rapid System Prototyping (RSP'04)*, pp. 80–87. IEEE Press, 2004.

67. Giacomazzi, P., and Trecordi, V., A study of non blocking multicast switching networks. *IEEE Transactions on Communications*, 43(2/3/4):1163–1168, February/March/April 1995.

68. Glesner, M., Hollstein, T., Indrusiak, L. S., Zipf, P., Pionteck, T., Petrov, M., Zimmer, H., and Murgan, T., Reconfigurable platforms for ubiquitous computing. In: *International Proceedings of the ACM Computing Frontiers Conference 2004; Ischia*, 2004.

69. Glynn, P. W. and Iglehart, D. L., Importance sampling for stochastic simulations. *Management Science*, 35:1367–1392, 1989.

70. Glynn, P. W. and Heidelberger, P., Analysis of parallel replicated simulations under a completion time constraint. *ACM Transactions on Modeling and Computer Simulation*, 1:3–23, 1991.

71. Goke, L.R. and Lipovski, G.J., Banyan networks for partitioning multiprocessor systems. *ACM SIGARCH Computer Architecture News*, 2(4):21–28, December 1973.

72. Gross, D. and Harris, C. M., *Fundamentals of Queueing Theory*. Wiley Interscience, 3rd ed., 1998.

73. Guerrier, P. and Grenier, A., A generic architecture for on-chip packet-switched interconnections. In: *Proceedings of IEEE Design Automation and Test in Europe (DATE 2000)*, pp. 250–256. IEEE Press, 2000.

74. Guo, M.-H. and Chang, R.-S., Multicast ATM switches: Survey and performance evaluation. *ACM Sigcomm: Computer Communication Review*, 28(2):98–131, April 1998.

75. Haas, P. J., *Stochastic Petri Nets*. Springer Verlag, 2002.

76. Halsall, F., *Data Communications, Computer Networks and Open Systems*. Addison Wesley Longman, London, 1996.

77. Händel, R., Huber, M. N., and Schröder, S., *ATM Networks*. Addison-Wesley, 2nd ed., 1995.

78. Hartenstein, R. W., Kress, R., and Reinig, H., A new FPGA architecture for word-oriented datapaths. In: *Proceedings of 4th International Workshop on Field-Programmable Logic and Applications (FPL '94)*, pp. 144–155, 1994.

79. Haverkort, B. R. and Trivedi, K.S., Specification and generation of Markov reward models. *Discrete-Event Dynamic Systems: Theory and Applications*, 3:219–247, 1993.

80. Heidelberger, P. and Lewis, P.A.W., Quantile estimation in dependent sequences. *Operations Research*, 32(1):185–209, 1984.

81. Heidelberger, P. and Welch, P.D., A spectral method for confidence interval generation and run length control in simulations. *Communications of the ACM*, 24(4):233–245, 1981.

82. Hellekalek, P., Good random number generators are (not so) easy to find. *Mathematics and Computers in Simulation*, pp. 485–505, 1998.

83. Hennessy, J. and Patterson, D. A., *Computer Architecture : A Quantitative Approach*. Morgan Kaufmann Publishers, San Mateo, 2nd ed., 2002.

84. Huang, T.-Y. and Wu, J.-L. C., Alternate resolution strategy in multistage interconnection networks. *Parallel Computing*, 20:887–896, 1994.

85. Jenq, Y.-C., Performance analysis of a packet switch based on single–buffered banyan network. *IEEE Journal on Selected Areas in Communications*, SAC–1(6):1014–1021, December 1983.

86. Jensen, K., *Coloured Petri Nets*. Springer Verlag, 2nd ed., 1996.

87. Jurczyk, M., Performance comparison of wormhole-routing priority switch architectures. In: *Proceedings International Conference on Parallel and Distributed Processing Techniques and Applications 2001 (PDPTA'01); Las Vegas*, pp. 1834–1840, 2001.

88. Kelling, C., *Simulationsverfahren für zeiterweiterte Petri-Netze*. SCS, 1995.

89. Kemeny, J. G. and Snell, J. L., *Finite Markov Chains*. The University Series in Undergraduate Mathematics. D. van Nostrand Company, Inc., Princeton, New Jersey, 1960.

90. Kendall, D.G., Some problems in the theory of queues. *Journal of the Royal Statistical Society, Series B*, 13:151–185, 1951.

91. Kermani, P. and Kleinrock, L., Virtual cut-through: A new computer communication switching technique. *Computer Neworks*, 3:267–286, 1979.

92. King, J., *Computer and Communication Systems Performance Modeling*. Prentice-Hall, Englewood Cliffs, N.J., 1990.

93. Kirschbaum, A. and Glesner, M., Rapid prototyping of communication architectures. In: *Proceedings of the 8th IEEE International Workshop on Rapid System Prototyping (IWRSP 1997)*, pp. 136–141. IEEE, 1997.

94. Kleinrock, L., Nomadicity: Anytime, anywhere in a disconnected world. *Mobile Networks and Applications*, 1(4):351–357, 1997.

95. Kleinrock, L., *Queueing Systems, Volume 1: Theory*. Wiley Interscience, New York, 1975.

96. Kleinrock, L., *Queueing Systems, Volume 2: Computer Applications*. Wiley Interscience, New York, 1976.

97. Knuth, D. E., *Art of Programming, Volume 2: Seminumerical Algorithms*. Addison-Wesley, 3rd ed., 1997.

98. Köhler, R.-D. and Kemmler, W., *Gigabit-Ethernet – 3COM – Die Komponenten der Zukunft*. Fossil-Verlag, 1999.

99. Koike, N., NEC Cenju-3: A microprocessor-based parallel computer. In: *Proceedings of the 8th International Symposium Parallel Processing*, pp. 396–401, April 1994.

100. Koppelman, D. M., Congested banyan network analysis using congested–queue states and neighboring–queue effects. *IEEE/ACM Transactions on Networking*, 4(1):106–111, February 1996.

101. Kouvatsos, D., Awan, I., and Al-Begain, K., Performance modelling of a wireless cell with multiple class services. *IEE Proceedings on Computers & Digital Technology*, 150(2):75–86, 2003.

102. Kruskal, C. P. and Snir, M., The performance of multistage interconnection networks for multiprocessors. *IEEE Transactions on Computers*, C–32(12):1091–1098, 1983.

103. Kruskal, C. P. and Snir, M., A unified theory of interconnection network structure. *Theoretical Computer Science*, 48(1):75–94, 1986.

104. Kruskal, C. P. and Snir, M., Optimal interconnection networks for parallel processors: The importance of being square. In: Yechiam Yemini, editor, *Current Advances in Distributed Computing and Communications*, pp. 91–113. Computer Science Press, Rockville, 1987.

105. Kumar, S., Jantsch, A., Soininen, J.-P., Forsell, M., Millberg, M., Öberg, J., Tiensyrjä, K., and Hemani, A., A network on chip architecture and design methodology. In: *Proceedings of the IEEE Computer Society Annual Symposium on VLSI (ISVLSI'02)*, pp. 105–112, 2002.

106. Lahiri, K., Dey, S., and Raghunathan, A., Evaluation of the traffic-performance characteristics of system-on-chip communication architectures. In: *Proceedings of the 14th International Conference on VLSI Design (VLSID '01)*, pp. 29–35. IEEE Press, 2001.

107. Law, A.M. and Kelton, W.D., *Simulation Modeling & Analysis*. McGraw-Hill, 3rd ed., 2000.

108. Lea, C.-T., Buffered or unbuffered: A case study based on $\log_d(n, e, p)$ networks. *IEEE Transactions on Communications*, 44(1):105–113, January 1996.

109. L'Ecuyer, P., Good parameters and implementations for combined multiple recursive random number generators. *Operations Research*, 47(1):159–164, January 1999.

110. Lee, G., Kang, B.-C., and Kain, R. Y., Analysis of finite buffered multistage combining networks. *IEEE Transactions on Parallel and Distributed Systems*, 6(7):760–766, July 1995.

111. Lee, J.-S. R., McNickle, D., and Pawlikowski, K., Quantile estimations in sequential steady-state simulation. In: *Proceedings of European Simulation Multiconference (ESM'99)*, pp. 168–174. SCS, 1999.

112. Lee, J.-S., Song, S.-J., Lee, K., Woo, J.-H., Kim, S.-E., Nam, B.-G., and Yoo, H.-J., An 800MHz star-connected on-chip network for application to systems on a chip. In: *Proceedings of 2003 IEEE International Solid-State Circuits Conference (ISSCC 2003)*, pp. 468–475. IEEE Press, 2003.

113. Lehoczky, J. P., Real-time queueing network theory. In: *Proceedings Real-Time Systems Symposium*, pp. 58–67. IEEE Computer Society Press, December 1997.

114. Leiserson, C. E., Fat-trees: Universal networks for hardware–efficient supercomputing. *IEEE Transactions on Computers*, 34(10):892–901, October 1985.

115. Leiserson, C. E., Abuhamdeh, Z. S., Douglas, D. C., Feynman, C. R., Ganmukhi, M. N., Hill, J. V., Hillis, W. D., Kuszmaul, B. C., Pierre, M. A. St., Wells, D. S., Wong-Chan, M. C., Yang, S.-W., and Zak, R., The network architecture of the Connection Machine CM-5. *Journal of Parallel and Distributed Computing*, 33:145–158, March 1996.

116. Leland, W. E., Taqqu, M. S., Willinger, W., and Wilson, D. V., On the self-similar nature of Ethernet traffic (extended version). *IEEE/ACM Transactions on Networking*, 2(1):1–14, February 1994.

117. Li, J. and Cheng, C.-K., Routability improvement using dynamic interconnect architecture. *IEEE Transactions on VLSI Systems*, 6(3):498–501, 1998.

118. Lindemann, C., *Performance Modelling with Deterministic and Stochastic Petri Nets*. John Wiley & Sons, 1998.

119. Lipovski, G. J. and Malek, M., *Parallel Computing: Theory and Comparisons*. John Wiley & Sons, New York, 1987.

120. Lu, W. and Giordano, S., Challenges in mobile ad hoc networking (collection of articles). *IEEE Communications Magazin*, 39(6), 2001.

121. Lucas, M. T., Dempsey, B. J., Wrege, D. E., and Weaver, A. C., (M,P,S) – an efficient background traffic model for wide-area network simulation. In: *Proceedings IEEE Global Telecommunications Conference*, volume 3, pp. 1572–1576. IEEE Computer Society, IEEE Computer Society Press, November 1997.

122. Lucas, M. T., Wrege, D. E., Dempsey, B. J., and Weaver, A. C., Statistical characterization of wide-area IP traffic. In: *Proceedings Sixth International Conference on Computer Communications and Networks*, pp. 442–447. IEEE Computer Society, IEEE Computer Society Press, September 1997.

123. Luciani, J. V. and Chen, C. Y. R., An analytical model for partially blocking finite–buffered switching networks. *IEEE/ACM Transactions on Networking*, 2(5):533–540, October 1994.

124. Lüdtke, D., Tutsch, D., Walter, A., and Hommel, G., Improved performance of bidirectional multistage interconnection networks by reconfiguration. In: *Proceedings of 2005 Design, Analysis, and Simulation of Distributed Systems (DASD 2005); San Diego*, pp. 21–27. SCS, April 2005.

125. Majer, M., Bobda, C., Ahmadinia, A., and Teich, J., Packet routing in dynamically changing networks on chip. In: *Proceedings of the 19th IEEE International Parallel and Distributed Processing Symposium (IPDPS 2005)*, page 154. IEEE Press, 2005.

126. Malek, M., The NOMADS republic. In: *Proceedings of International Conference on Advances in Infrastructure for Electronic Business, Education, Science, Medicine and Mobile Technologies on the Internet; L'Aquila*. Scuola Superiore G. Reiss Romoli (SSGRR), Telecom Italia, 2003.

127. Maltz, D., Broch, J., and Johnson, D., Lessons from a full-scale multihop wireless ad hoc network testbed. *IEEE Personal Communications*, 8(1), 2001.

128. Marsan, M. A., Stochastic Petri nets: An elementary introduction. In: G. Rozenberg, editor, *Advances in Petri Nets*, volume 424 of *Lecture Notes in Computer Science*, pp. 1–29. Springer Verlag, Berlin, 1990.

129. Marsan, M. A., Balbo, G., and Conte, G., A class of generalized stochastic Petri nets for the performance evaluation of multiprocessor systems. *ACM Transactions on Computer Systems*, 2(2):93–122, May 1984.

130. Marsan, M. A. and Chiola, G., On Petri nets with deterministic and exponential transition firing times. In: G. Rozenberg, editor, *Proceedings of the 7th European Workshop on Application and Theory of Petri Nets; Oxford*, volume 266 of *Advances in Petri Nets 1987, Lecture Notes on Computer Science*. Springer Verlag, Berlin, 1986.

131. Marsan, M. A., Marano, S., Mastroianni, C., and Meo, M., Performance analysis of cellular mobile communication networks supporting multimedia services. *Mobile Networks and Applications*, 5:167–177, 2000.

132. Marsan, M. A., Chiasserini, C.-F., and Fumagalli, A., Dimensioning handover buffers in wireless ATM networks with GSPN models. In: *19th Int. Conf. on Application and Theory of Petri Nets (ICATPN'98)*, pp. 44–63. Springer-Verlag, 1998.

133. Marsan, M. A. and Gaeta, R., Modeling ATM systems with GSPNs and SWNs. *ACM Sigmetrics: Performance Evaluation Review*, 2(26):28–37, August 1998.

134. Matsumoto, M., and Nishimura, T., Mersenne twister: a 623-dimensionally equidistributed uniform pseudo-random number generator. *ACM Transactions on Modeling and Computer Simulation*, 8(1):3–30, 1998.

135. Maxfield, C., *The Design Warrior's Guide to FPGAs*. Newnes / Elsevier, 2004.

136. McKinley, P. K., Tsai, Y.-J., and Robinson, D. F., Collective communication in wormhole–routed massively parallel computers. *Computer*, 28(12):39–50, December 1995.

137. Medhi, J., *Stochastic Models in Queueing Theory*. Academic Press, 2002.
138. Meketon, M.S. and Schmeister, B., Overlapping batch means: Something for nothing. In: *Proceedings of the 1984 Winter Simulation Conference*, pp. 227–230. SCS, 1984.
139. Mohapatra, P., Wormhole routing techniques for directly connected multicomputer systems. *ACM Computing Surveys*, 30(3):374–410, September 1998.
140. Mohapatra, P. and Das, C. R., Performance analysis of finite–buffered asynchronous multistage interconnection networks. *IEEE Transactions on Parallel and Distributed Systems*, 7(1):18–25, January 1996.
141. Molloy, M.K., *On the Integration of Delay and Throughput Measures in Distributed Processing Models*. PhD thesis, University of California Los Angeles, 1981.
142. Mun, Y. and Youn, H. Y., Performance analysis of finite buffered multistage interconnection networks. *IEEE Transactions on Computers*, 43(2):153–161, February 1994.
143. Murali, S. and De Micheli, G., SUNMAP: A tool for automatic topology selection and generation for NoCs. In: *Proceedings of the 41st Design Automation Conference (DAC 2004)*, pp. 914–919. ACM, 2004.
144. Murata, T., Petri nets: Properties, analysis and applications. *Proceedings of the IEEE*, 77(4):541–580, April 1989.
145. Narasimha, M., The Batcher-Banyan self-routing network: Universality and simplification. *IEEE Transactions on Communications*, 36(10):1175–1178, October 1988.
146. Natkin, S., *Les Reseaux de Petri stochastiques et leur Application à l'Evaluation des Systèmes Informatiques*. PhD thesis, CNAM Paris, 1980.
147. Nelson, B.L., A perspective on variance reduction in dynamic simulation experiments. *Communications in Statistics (B: Simulation and Computation)*, 16(2):385–426, 1987.
148. Ni, L. M., Gui, Y., and Moore, S., Performance evaluation of switch-based wormhole networks. *IEEE Transactions on Parallel and Distributed Systems*, 8(5):462–474, May 1997.
149. Ortega, J. M. and Rheinboldt, W. C., *Iterative Solution of Nonlinear Equations in Several Variables*. Computer Science and Applied Mathematics. Academic Press, New York, 1970.
150. Pahlavan, K. and Krishnamurthy, P., *Principles of Wireless Network*. Prentice Hall, 2002.
151. Park, J. and Yoon, H., Cost-effective algorithms for multicast connection in ATM switches based on self-routing multistage networks. *Computer Communications*, 21:54–64, 1998.
152. Patel, J. H., Performance of processor–memory interconnections for multiprocessors. *IEEE Transactions on Computers*, C–30(10):771–780, October 1981.
153. Pawlikowski, K., Steady-state simulation of queueing processes: A survey of problems and solutions. *ACM Computing Surveys*, 22(2):123–170, 1990.
154. Pawlikowski, K., Jeong, H.-D. J., and Lee, J.-S. R., On credibility of simulation studies of telecommunication networks. *IEEE Communications Magazine*, pp. 132–139, January 2002.
155. Pawlikowski, K., Yau, V. W. C., and McNickle, D., Distributed stochastic discrete-event simulation in parallel time streams. In: *Proceedings of the 1994 Winter Simulation Conference; Lake Buena Vista*, pp. 723–730, December 1994.

156. Pelz, E. and Tutsch, D., Modeling multistage interconnection networks of arbitrary crossbar size with compositional high level Petri nets. In: *Proceedings of the 2005 European Simulation and Modelling Conference (ESM 2005); Porto*, pp. 537–543. Eurosis, 2005.

157. Perkins, C., et al., IP mobility support. RFC 2002, IETF, 1996.

158. Perkins, C., *Ad Hoc Networking*. Addison Wesley, 2001.

159. Peterson, R., Ziemer, R., and Borth, D., *Introduction to Spread Spectrum Communications*. Prentice Hall, 1995.

160. Petri, C. A., *Kommunikation mit Automaten*. PhD thesis, Universität Bonn, 1962.

161. Ponnekanti, S. R., Lee, B., Fox, A. Hanrahan, P., and Winograd, T., ICrafter: A service framework for ubiquitous computing environments. In: *Proceedings of the Ubiquitous Computing Conference (UBICOMP 2001)*. Springer Verlag, 2001.

162. Raguin, D., Kubisch, M., Karl, H., and Wolisz, A., Queue-driven cut-through medium access in wireless ad hoc networks. In: *Proceedings of IEEE Wireless Communications and Networking Conference (WCNC'04); Atlanta*. IEEE, 2004.

163. Rauber, T. and Rünger, G., *Parallele und verteilte Programmierung*. Springer Verlag, Berlin, 2000.

164. Reibman, A. and Trivedi, K., Numerical transient analysis of markov models. *Computers and Operations Research*, 15(1):19–36, 1988.

165. Reisig, W., *Petri Nets: an Introduction*. Springer Verlag, Heidelberg, 1985.

166. Ren, W., Siu, K.-Y., Suzuki, H., and Shinohara, M., Multipoint-to-multipoint ABR service in ATM. *Computer Networks and ISDN Systems*, 30:1793–1810, 1998.

167. Rettberg, R.D., Crowther, W.R., Carvey, P.P., and Tomlinson, R.S., The Monarch parallel processor hardware design. *Computer*, 23(4):18–30, April 1990.

168. Robinson, D. F., Judd, D., McKinley, P. K., and Cheng, B. H. C., Efficient multicast in all–port wormhole–routed hypercubes. *Journal of Parallel and Distributed Computing*, 31(2):126–140, December 1995.

169. Rohatgi, V.K., *An Introduction to Probability Theory and Mathematical Statistics*. John Wiley & Sons, 1976.

170. Rosen et al., Multiprotocol label switching architecture. Internet draft, IETF, 1999.

171. Rădulescu, A. and Goossens, K., Communication services for networks on chip. In: Shuvra S. Bhattacharyya, Ed F. Deprettere, and Jїgen Teich, editors, *Domain-Specific Processors: Systems, Architectures, Modeling, and Simulation*, pp. 193–213. Marcel Dekker, 2004.

172. Rubinstein, R.Y., *Simulation and the Monte-Carlo-Method*. John Wiley & Sons, 1981.

173. Sánchez, J. L. and García, J. M., Dynamic reconfiguration of node location in wormhole networks. *Journal of Systems Architecture*, 46(10):873–888, 2000.

174. Schiller, J., *Mobile Communications*. Addison Wesley, London, 2nd ed., 2003.

175. Schruben, L.W., Singh, H., and Tierney, L., Optimal test for initialization bias in simulation output. *Operations Research*, 31(6):1167–1178, 1983.

176. Shaikh, S. Z., Schwartz, M. and Szymanski, T. H., Analysis, control and design of crossbar and banyan based broadband packet switches for integrated traffic. In: *IEEE International Conference On Communications ICC'90 Including Supercomm Technical Sessions. SUPERCOMM ICC'90 Conference Record; Atlanta*, volume 2, pp. 761–765, New York, April 1990.

177. Sharma, N. K., Review of recent shared memory based ATM switches. *Computer Communications*, 22:297–316, 1999.

178. Sherman, M. and Carlstein, E., Confidence intervals based on estimators with unknown rates of convergence. *Computational Statistics and Data Analysis*, pp. 123–139, 2004.

179. Shi, H. and Sethu, H., Virtual circuit blocking probabilities in an ATM banyan network with bxb switching elements. In: *Proceedings of the Applied Telecommunication Symposium 2001 (ATS'01); Seattle*, pp. 21–25. SCS, 2001.

180. Sibal, S. and Zhang, J., On a class of banyan networks and tandem banyan switching fabrics. *IEEE Transactions on Communications*, 43(7):2231–2240, July 1995.

181. Sima, D., Fountain, T., and Kacsuk, P., *Advanced Computer Architectures*. Addison Wesley, 1997.

182. Sivaram, R., Panda, D.K., and Stunkel, C.B., Efficient broadcast and multicast on multistage interconnection networks using multiport encoding. *IEEE Transaction on Parallel and Distributed Systems*, 9(10):1004–1028, October 1998.

183. Sokol, J. and Widmer, J., USAIA Ubiquitous Service Access Internet Architecture. Technical Report TR–01–003, International Computer Science Institute, Berkeley, 2001.

184. Soumiya, T., Nakamichi, K., Kakuma, S., Hatano, T., and Hakata, A., The large capacity ATM backbone switch "FETEX-150 ESP". *Computer Networks*, 31(6):603–615, 1999.

185. Stallings, W., *Wireless Communications and Networking*. Prentice Hall, 2002.

186. Stallings, W., *High-Speed Networks*. Prentice Hall, New Jersey, 1998.

187. Stergiou, S., Angiolini, F., Carta, S., Raffo, L., Bertozzi, D., and De Micheli, G., xpipes lite: A synthesis oriented design library for networks on chips. In: *Proceedings of the Design, Automation and Test in Europe Conference and Exhibition (DATE'05)*, volume 2, pp. 1188–1193. IEEE, 2005.

188. Stevens, W.R., *TCP/IP Illustrated, Volume 1 – The Protocols*. Addison-Wesley, Boston, 1994.

189. Stewart, W., *Introduction to Numerical Solution of Markov Chains*. Princeton University Press, Princeton, N.J., 1994.

190. Stüber, G.L., *Principles of Mobile Communication*. Kluwer Academic Publishers, Boston, 2nd ed., 2001.

191. Stunkel, C.B., Shea, D.G., Abali, B., Atkins, M.G., Bender, C.A., Grice, D.G., Hochschild, P., Joseph, D.J., Nathanson, B.J., Swetz, R.A., Stucke, R.F., Tsao, M., and Varker, P.R., The SP2 high-performance switch. *IBM Systems Journal*, 34(2):185–204, 1995.

192. Subramaniam, S. and Somani, A.K., Multicasting in ATM networks using MINs. *Computer Communications*, 19:712–722, 1996.

193. Takahashi, Y., A lumping method for numerical calculations of stationary distributions of Markov chains. Research report B 18, Tokyo Institute of Technology, Department of Information Science, Tokyo, 1975.

194. Tanenbaum, A.S., *Modern Operating Systems*. Prentice Hall, Englewood Cliffs, N.J., 2nd ed., 2001.

195. Tanenbaum, A.S., *Computer Networks*. Prentice Hall, 4th ed., 2002.

196. Tanenbaum, A.S. and van Steen, M., *Distributed Systems: Principles and Paradigms*. Prentice Hall, 1st ed., 2002.

197. Tobagi, F.A., Kwok, T., and Chiussi, F.M., Architecture, performance, and implementation of the tandem-banyan fast packet switch. *IEEE Journal on Selected Areas of Communication*, 9(8):1173–1193, October 1991.

198. Trimberger, S., Carberry, D., Johnson, A., and Wong, J., A time-multiplexed FPGA. In: *Proceedings of the IEEE Symposium on Field-Programmable Custom Computing Machines*, pp. 22–28, 1997.

199. Trivedi, K., *Probability and Statistics with Reliability, Queuing, and Computer Science Applications*. Prentice-Hall, Englewood Cliffs, N.J., 1982.

200. Turner, J. and Melen, R., Multirate Clos networks. *IEEE Communications Magazine*, 41(10):38–44, October 2003.

201. Tutsch, D., Model comparison by performance evaluation of multistage interconnection networks. In: *Proceedings of the High Performance Computing Symposium 1998 (HPC'98); Boston*, pp. 369–374. SCS, April 1998.

202. Tutsch, D., Object-oriented modeling of interconnection networks. In: *Proceedings of the International Workshop on Communication Based Systems (CBS'98); Berlin*, pp. 107–116, October 1998.

203. Tutsch, D., *Verfahren zur Leistungsbewertung von gepufferten mehrstufigen Verbindungsnetzwerken*. PhD thesis, Technische Universität Berlin, 1998.

204. Tutsch, D., Performance analysis of transient network behavior in case of packet multicasting. In: *Proceedings of the European Simulation Symposium 1999 (ESS'99); Erlangen*, pp. 630–634. SCS, October 1999.

205. Tutsch, D., Transient multicast traffic performance of MINs: A case study. In: *Workshop Distributed Computing on the Web 1999 (DCW'99); Rostock*, pp. 103–110. GI, June 1999.

206. Tutsch, D., Buffer design in delta networks. In: Günter Hommel and Sheng Huanye, editors, *The Internet Challenge: Technology and Applications*, pp. 93–101. Kluwer Academic Publishers, 2002.

207. Tutsch, D. and Brenner, M., Multicast probabilities of multistage interconnection networks. In: *Proceedings of the 12th European Simulation Symposium 2000 (ESS'00); Hamburg*, pp. 554–558. SCS, September 2000.

208. Tutsch, D. and Brenner, M., MINSimulate – a multistage interconnection network simulator. In: *17th European Simulation Multiconference: Foundations for Successful Modelling & Simulation (ESM'03); Nottingham*, pp. 211–216. SCS, June 2003.

209. Tutsch, D., Brenner, M., and Hommel, G., Performance analysis of multistage interconnection networks in case of cut-through switching and multicasting. In: *Proceedings of the High Performance Computing Symposium 2000 (HPC 2000); Washington DC*, pp. 377–382. SCS, April 2000.

210. Tutsch, D., Hendler, M., and Hommel, G., Multicast performance of multistage interconnection networks with shared buffering. In: *Proceedings of the IEEE International Conference on Networking (ICN 2001); Colmar*, pp. 478–487. IEEE, July 2001.

211. Tutsch, D. and Holl-Biniasz, R., Performance evaluation using measure dependent transitions in Petri nets. In: *Proceedings of the Fifth International Symposium on Modeling, Analysis and Simulation of Computer and Telecommunication Systems (MASCOTS'97); Haifa*, pp. 235–240. IEEE Computer Society Press, January 1997.

212. Tutsch, D. and Hommel, G., Performance of buffered multistage interconnection networks in case of packet multicasting. In: *Proceedings of the 1997 Conference on Advances in Parallel and Distributed Computing (APDC'97); Shanghai*, pp. 50–57. IEEE Computer Society Press, March 1997.

213. Tutsch, D. and Hommel, G., Multicasting in buffered multistage interconnection networks: an analytical algorithm. In: *12th European Simulation Multiconference: Simulation – Past, Present and Future (ESM'98); Manchester*, pp. 736–740. SCS, June 1998.

214. Tutsch, D. and Hommel, G., Multicasting in interconnection networks: Modeling and performance evaluation. In: *Proceedings of the High Performance Computing Symposium 1999 (HPC'99); San Diego*, pp. 413–424. SCS, April 1999.

215. Tutsch, D. and Hommel, G., Multifractal multicast traffic in multistage interconnection networks. In: *Proceedings of the High Performance Computing Symposium 2001 (HPC 2001); Seattle*, pp. 257–262. SCS, April 2001.

216. Tutsch, D. and Hommel, G., Comparing switch and buffer sizes of multistage interconnection networks in case of multicast traffic. In: *Proceedings of the High Performance Computing Symposium 2002 (HPC 2002); San Diego*, pp. 300–305. SCS, April 2002.

217. Tutsch, D. and Hommel, G., Generating systems of equations for performance evaluation of buffered multistage interconnection networks. *Journal of Parallel and Distributed Computing*, 62(2):228–240, February 2002.

218. Tutsch, D. and Hommel, G., Multilayer multistage interconnection networks. In: *Proceedings of 2003 Design, Analysis, and Simulation of Distributed Systems (DASD 2003); Orlando*, pp. 155–162. SCS, April 2003.

219. Tutsch, D. and Hommel, G., Multicast routing in Clos networks. In: *Proceedings of 2004 Design, Analysis, and Simulation of Distributed Systems (DASD 2004); Arlington*, pp. 21–27. SCS, April 2004.

220. Tutsch, D. and Lüdtke, D., Multicast in switches: Packet sequences versus independent packets. In: *Proceedings of 2005 Design, Analysis, and Simulation of Distributed Systems (DASD 2005); San Diego*, pp. 46–52. SCS, April 2005.

221. Tutsch, D., Lüdtke, D., and Kühm, M., Investigating dynamic reconfiguration of network architectures with CINSim. In: *Proceedings of the 13th Conference on Measurement, Modeling, and Evaluation of Computer and Communication Systems 2006 (MMB 2006); Nürnberg*, pp. 445–448. VDE, March 2006.

222. Tutsch, D., Lüdtke, D., Walter, A., and Kühm, M., CINSim – a component-based interconnection network simulator for modeling dynamic reconfiguration. In: *Proceedings of the 12th International Conference on Analytical and Stochastic Modelling Techniques and Applications (ASMTA 2005); Riga*, pp. 132–137. IEEE/SCS, June 2005.

223. Tutsch, D. and Sokol, J., Petri net based performance evaluation of USAIA's bandwidth partitioning for the wireless cell level. In: *9th International Workshop on Petri Nets and Performance Models (PNPM'01); Aachen*, pp. 49–58. IEEE Computer Society, 2001.

224. Vaidyanathan, R. and Trahan, J.L., *Dynamic Reconfiguration: Architectures and Algorithms*. Kluwer Academic, New York, 2003.

225. Varavithya, V. and Mohapatra, P., Asynchronous tree-based multicasting in wormhole-switched multistage interconnection networks. *IEEE Transactions on Parallel and Distributed Systems*, pp. 1159–1178, November 1999.

226. Villen-Altamirano, M. and Villen-Altamirano, J., RESTART: A methode for accelerating rare event simulations. In: J.W. Cohen and C.D. Pack, editors, *Queueing Performance and Control in ATM (ITC-13)*, pp. 71–76. Elsevier Science Publishers, North-Holland, 1991.

227. Waugh, T.C., Field programmable gate array: Key to reconfigurable array out-performance supercomputers. In: *Proceedings of the Custom Integrated Circuits Conference 1991*, pp. 6.6/1–4, 1991.

228. Weiser, M., Some computer science issues in ubiquitous computing. *Communications ACM*, 36(7):74–84, 1993.

229. Wesel, E., *Wireless Multimedia Communications: Networking Video, Voice, and Data*. Addison Wesley, London, 1998.

230. Widjaja, I., Leon-Garcia, A., and Mouftah, H.T., The effect of cut-through switching on the performance of buffered banyan networks. *Computer Networks and ISDN Systems*, 26:139–159, 1993.

231. Wiklund, D. and Liu, D., SoCBUS: Switched network on chip for hard real time embedded systems. In: *Proceedings of the 17th IEEE International Parallel and Distributed Processing Symposium (IPDPS 2003)*, pp. 78–85. IEEE Press, 2003.

232. Willinger, W., Taqqu, M.S., Sherrman, R., and Wilson, D.V., Self-similarity through high-variablility: Statistical analysis of Ethernet LAN traffic at the source level. *IEEE/ACM Transactions on Networking*, 5(1):71–86, 1997.

233. Wilson, J.R., Variance reduction techniques for digital simulation. *American Journal on Mathematics in Management Science*, 4(3):277–312, 1984.

234. Wingard, D., MicroNetwork-based integration for SoCs. In: *Proceedings of the Design Automation Conference (DAC 2001)*, pp. 673–677. ACM, 2001.

235. Wolf, T. and Turner, J., Design issues for high performance active routers. *IEEE Journal on Selected Areas of Communications*, 19(3):404–409, March 2001.

236. Wolisz, A., Wireless internet architectures: Selected issues. In: J. Wozniak and J. Konorski, editors, *Personal Wireless Communications*, pp. 1–16. Kluwer, 2000.

237. Wong, P.C. and Yeung, M.S., Design and analysis of a novel fast packet switch–pipeline banyan. *IEEE/ACM Transactions on Networking*, 3(1):63–69, February 1995.

238. Xiong, Y. and Mason, L., Analysis of multicast ATM switching networks using CRWR scheme. *Computer Networks and ISDN Systems*, 30:835–854, 1998.

239. Xu, H., Gui, Y., and Ni, L.M., Optimal software multicast in wormhole-routed multistage networks. *IEEE Transactions on Parallel and Distributed Systems*, 8(6):597–606, June 1997.

240. Yang, Y., An analytical model for the performance of buffered multicast banyan networks. *Computer Communications*, 22:598–607, 1999.

241. Yang, Y., A class of multistage conference switching networks for group communication. In: *Proc. 2002 International Conference on Parallel Processing (ICPP 2002); Vancouver*, pp. 73–80. IEEE, August 2002.

242. Yang, Y. and Wang, J., A class of multistage conference switching networks for group communication. *IEEE Transactions on Parallel and Distributed Systems*, 15(3):228–243, March 2004.

243. Yasuda, Y., Fujii, H., Akashi, H., Inagami, Y., Tanaka, T., Wada, H., and Sumimoto, T., Deadlock-free fault-tolerant routing in the multi-dimensional crossbar network and its implementation for the Hitachi SR2201. In: *Proceedings of the 11th International Parallel Processing Symposium (IPPS'97); Genf*, pp. 346–352. IEEE Computer Society Press, April 1997.

244. Yoon, H., Lee, K.Y., and Liu, M.T., Performance analysis of multibuffered packet–switching networks in multiprocessor systems. *IEEE Transactions on Computers*, 39(3):319–327, March 1990.

245. Youn, H.Y. and Chen, C. C.-Y., A comprehensive performance evaluation of crossbar networks. *IEEE Transactions on Parallel and Distributed Systems*, 4(5):481–489, May 1993.

246. Youn, H.Y. and Mun, Y., On multistage interconnection networks with small clock cycles. *IEEE Transactions on Parallel and Distributed Systems*, 6(1):86–93, January 1995.

247. Yuen, S., Kropf, P.G., Unger, H., and Babin, G., Simulation of communities of nodes in a wide area distributed system. In: *Proceedings of the EUROSIM 2001; Delft*. IEEE, 2001.

248. Zeigler, B. P., Praehofer, H., and Kim, T. G., *Theory of Modeling and Simulation*. Academic Press, 2nd ed., 2000.

249. Zhou, B. and Atiquzzaman, M., Efficient analysis of multistage interconnection networks using finite output-buffered switching elements. *Computer Networks and ISDN Systems*, 28:1809–1829, 1996.

250. Zhou, B. and Atiquzzaman, M., A performance comparison of four buffering schemes for multistage interconnection networks. *International Journal of Parallel and Distributed Systems and Networks*, 5(1):17–25, 2002.

# Index

Neil J. Gunther, Performance Dynamics Company,
Castro Valley, CA, USA

# Analyzing Computer System Performance with Perl::PDQ

XXIII, 436 p. 176 illus. Hardcover
ISBN 3-540-20865-8

Analyzing computer system performance is often regarded by most system administrators, IT professionals and software engineers as a black art that is too time consuming to learn and apply. Finally, this book by acclaimed performance analyst Dr. Neil Gunther makes this subject understandable and applicable through programmatic examples. The means to this end is the open-source performance analyzer Pretty Damn Quick (PDQ) written in Perl and available for download from www.perfdynamics.com.

As the epigraph in this book points out, Common sense is the pitfall of performance analysis. The performance analysis framework that replaces common sense is revealed in the first few chapters of Part I. The important queueing concepts embedded in PDQ are explained in a very simple style that does not require any knowledge of formal probability theory. Part II begins with a full specification of how to set up and use PDQ replete with examples written in Perl. Subsequent chapters present applications of PDQ to the performance analysis of multicomputer architectures, benchmark results, client/server scalability, and Web-based applications. The examples are not mere academic toys but are based on the author's experience analyzing the performance of large-scale systems over the past 20 years. By following his lead, you will quickly be able to set up your own Perl scripts for collecting data and exploring performance-by-design alternatives without inflating your manager's schedule.

**Contents:** Part I System Theory: Time - The Zeroth Performance Metric; Getting the Jump on Queueing; Queueing Systems for Computer Systems; Linux Load Average - Take a Load Off!; Performance Bounds and Log Jams.- Part II System Practice: Pretty Damn Quick: A Slow Introduction; Analyzing Multicomputer Architectures; How to Measure an Elephant with PDQ; Analyzing Client/Server Applications; Analyzing Web Applications with PDQ.- Part III Appendices: Glossary of Terms; A Short History of Buffers; Thanks for the (Lack of) Memories; Performance Metrics and Tools; List of Programs; Compendium of Queueing Equations; Solutions to Selected Exercises.